DNA METHYLATION

Approaches, Methods, and Applications

DNA METHYLATION

Approaches, Methods, and Applications

EDITED BY

Manel Esteller

CRC PRESS

Boca Raton   London   New York   Washington, D.C.

Library of Congress Cataloging-in-Publication Data

DNA methylation: approaches, methods and applications / edited by Manel Esteller.
    p. cm.
Includes bibliographical references and index.
ISBN 0-8493-2050-X (alk. paper)
1. DNA--Methylation--Handbooks, manuals, etc. I. Esteller, Manel. II. Title.

QP624.5.M46D626 2004
616'.042—dc22   2004049665

Visit the CRC Press Web site at www.crcpress.com

© 2005 by CRC Press LLC

No claim to original U.S. Government works
International Standard Book Number 0-8493-2050-X
Library of Congress Card Number 2004049665
Printed in the United States of America 1 2 3 4 5 6 7 8 9 0
Printed on acid-free paper

Preface

A quick glance at the number of references to DNA methylation in the searchable biomedical databases, such as PubMed and others, or the frequency with which these words appear on Internet websites and public databases demonstrates the current boom in interest in DNA methylation. This book is born of the necessity to serve that interest: it aims to be a comprehensive handbook covering the areas in which DNA methylation is important, addressing how this modification may be studied for academic and clinicopathological purposes and how it may be applied in therapeutic strategies. Until now, the field has lacked this type of organized bibliographical source, but this current book, with contributions from many of the leading experts in the field of DNA methylation, makes good this absence.

Editor

Manel Esteller graduated with honors in medicine from the University of Barcelona, Catalonia, Spain, and then earned his Ph.D. *cum laude* from the University of Barcelona. Dr. Esteller's credits include the following: invited researcher at the School of Biological and Medical Sciences, University of St. Andrews, Scotland, and postdoctoral researcher and research associate at The Johns Hopkins University and School of Medicine, Baltimore, Maryland. He is currently director of the Cancer Epigenetics Branch at the Spanish National Cancer Centre (CNIO) in Madrid. Through his efforts, the hypermethylation-associated silencing of tumor suppressor genes is in the forefront not only of epigenetics, but also current cancer research.

The author of more than ninety original peer-reviewed manuscripts in biomedical sciences, Dr. Esteller is a member of eight international scientific societies. He has served as a reviewer for many journals and funding agencies in the biomedical area. His numerous awards include Best Young Investigator, awarded by the European Association for Cancer Research (2000), Best Young Cancer Researcher, awarded by the European School of Medical Oncology (1999), and First Prize in Basic Research at the Johns Hopkins Oncology Center (1999).

Contributors

Esteban Ballestar
Cancer Epigenetics Laboratory
Spanish National Cancer Centre
Madrid, Spain

Robert Brown
Department of Medical Oncology
Cancer Research UK, Beatson
  Laboratories
Glasgow University
Glasgow, Scotland

Michael W.-Y. Chan
Department of Molecular Virology,
  Immunology, and Medical Genetics
Comprehensive Cancer Center
The Ohio State University
Columbus, Ohio

Jonathan C. Cheng
Department of Urology
University of Southern
  California/Norris Comprehensive
  Cancer Center
Keck School of Medicine
Los Angeles, California

Susan J. Clark
Sydney Cancer Centre
Kanematsu Laboratories
Royal Prince Alfred Hospital
Camperdown, Australia

Rainer Claus
Department of Hematology
University of Freiburg Medical Center
Freiburg, Germany

Joseph F. Costello
Department of Neurological Surgery
  and The Brain Tumor Research Center
University of California
San Francisco, California

Maurizio D'Esposito
Institute of Genetics and Biophysics
  "Adriano Buzzati Traverso"
Consiglio Nazionale delle Ricerche
Naples, Italy

Jesus Espada
Cancer Epigenetics Laboratory
Spanish National Cancer Centre
Madrid, Spain

Manel Esteller
Cancer Epigenetics Laboratory
Spanish National Cancer Centre
Madrid, Spain

Mario F. Fraga
Cancer Epigenetics Laboratory
Spanish National Cancer Centre
Madrid, Spain

Jordi Frigola
Institut de Recerca Oncologica
Hospital Duran i Reynals
L'Hospitalet
Barcelona, Spain

Jens Hasskarl
Department of Hematology
University of Freiburg Medical Center
Freiburg, Germany

Rui Henrique
Department of Otolaryngology —
Head and Neck Surgery
The Johns Hopkins School of Medicine
Baltimore, Maryland

James G. Herman
The Sidney Kimmel Comprehensive
Cancer Center
Johns Hopkins School of Medicine
Baltimore, Maryland

Tim H.-M. Huang
Department of Molecular Virology,
Immunology, and Medical Genetics
Comprehensive Cancer Center
The Ohio State University
Columbus, Ohio

Carmen Jerónimo
Department of Otolaryngology —
Head and Neck Surgery
The Johns Hopkins School of Medicine
Baltimore, Maryland

Peter A. Jones
Department of Urology
University of Southern
California/Norris Comprehensive
Cancer Center
Keck School of Medicine
Los Angeles, California

Peter W. Laird
Departments of Surgery and of
Biochemistry and Molecular Biology
University of Southern
California/Norris Comprehensive
Cancer Center
Los Angeles, California

Michael Lübbert
Department of Hematology
University of Freiburg Medical Center
Freiburg, Germany

Maria Rosaria Matarazzo
Institute of Genetics and Biophysics
"Adriano Buzzati Traverso"
Consiglio Nazionale delle Ricerche
Naples, Italy

Maria F. Paz
Cancer Epigenetics Laboratory
Spanish National Cancer Centre
Madrid, Spain

Miguel A. Peinado
Institut de Recerca Oncologica
Hospital Duran i Reynals
L'Hospitalet
Barcelona, Spain

Christoph Plass
Division of Human Cancer Genetics
Medical Research Facility
The Ohio State University
Columbus, Ohio

Laura J. Rush
Department of Veterinary Biosciences
The Ohio State University
Columbus, Ohio

Masahiko Shiraishi
Department of Molecular Virology,
Immunology, and Medical Genetics
Comprehensive Cancer Center
The Ohio State University
Columbus, Ohio

David Sidransky
Department of Otolaryngology —
Head and Neck Surgery
The Johns Hopkins School of Medicine
Baltimore, Maryland

Dominic J. Smiraglia
Department of Cancer Genetics
Roswell Park Cancer Institute
Buffalo, New York

Nicole M. Sodir
Departments of Surgery and of
  Biochemistry and Molecular Biology
University of Southern
  California/Norris Comprehensive
  Cancer Center
Los Angeles, California

Maria Strazzullo
Institute of Genetics and Biophysics
  "Adriano Buzzati Traverso"
Consiglio Nazionale delle Ricerche
Naples, Italy

Daniel J. Weisenberger
Department of Urology
University of Southern
  California/Norris Comprehensive
  Cancer Center
Keck School of Medicine
Los Angeles, California

Table of Contents

1 Impact of DNA Methylation on Health and Disease

Manel Esteller

CONTENTS

THE RELEVANCE OF DNA METHYLATION TO HUMAN BIOLOGY, CLINICOPATHOLOGICAL SYNDROMES, AND EXPERIMENTAL MODELS

DNA METHYLATION IN PHYSIOLOGICAL CONDITIONS

We can lump within the scope of the enigmatic word of "epigenetics" all the heritable changes in gene expression patterns that are based on factors other than straightforward DNA sequences. The mechanisms controlling epigenetics are complex and we have only just begun to get our first glimpses of their nature. For example, chromatin structure, controlled by the patterns of acetylation and methylation of the histone proteins around the regulatory regions of genes (Jenuwein and Allis, 2001), is one critical layer of epigenetics [1]. However, at a deeper level still, the most "genetic" of all epigenetic modifications is DNA methylation [2].

In humans, the vast majority of DNA methylation occurs in the cytosine of the CpG dinucleotides. We need certain levels of methylcytosine in our genomes to be considered normal human beings. Endoparasitic sequences such as Alu elements or LINEs (long interspersed nuclear elements) refrain from jumping around thanks to the repression of DNA methylation; we keep our parental marks in our imprinted genes thanks to DNA methylation; and the need to silence one of each pair of X chromosomes in women is met by DNA methylation. At the same time, the distribution of the dinucleotide CpG in our genomes is not random. Most ubiquitously expressed genes have high concentrations of CpGs in their promoter-regulatory regions, from where the RNA transcript of the gene originates. These regions are called CpG islands. In a normal cell, they are unmethylated, and the gene is expressed if the required transcription factors are present.

DNA METHYLATION IN CANCER

The perfect epigenetic equilibrium of DNA methylation previously described in the normal cell is dramatically transformed in the cancer cell. The DNA methylation aberrations observed can be considered as falling into one of two categories: transcriptional silencing of tumor suppressor genes by CpG island promoter hypermethylation and a massive global genomic hypomethylation. Let us briefly examine these two features.

Hypermethylation of Tumor Suppressor Genes

In human tumors, some CpG islands become hypermethylated with the result that the expression of the contiguous gene is shut down. If this aberration affects a tumor suppressor gene, it confers a selective advantage on that cell and is selected generation after generation. We and other researchers have contributed to the identification of a long list of hypermethylated genes in human neoplasias [3,4], and this epigenetic alteration is now considered to be a common hallmark of all human cancers affecting all cellular pathways [5–7]. Extremely important genes in cancer biology, such as the cell-cycle inhibitor $p16^{INK4a}$; the p53-regulator $p14^{ARF}$; the DNA-repair genes hMLH1, BRCA1, and MGMT; the cell-adherence gene E-cadherin; and the estrogen and retinoid receptors, undergo methylation-associated silencing in cancer cells [6].

The profiles of CpG island hypermethylation are known to depend on the tumor type [3,4]. Each tumor subtype can now be assigned a CpG island hypermethylation profile (methylotype) that almost completely defines that particular malignancy in a similar fashion, as do genetic and cytogenetic markers. Establishing a DNA methyl-fingerprint can be very useful for classifying these malignancies according to their aggressiveness or sensitivities to chemotherapy. Single-gene approaches can also be extremely useful. In gliomas, B-diffuse large cell lymphomas for example, we have demonstrated that hypermethylation of the DNA repair gene MGMT confers a good response to the chemotherapy regimens that include the alkylating drugs BCNU and cyclophosphamide [8–10].

Using DNA Hypermethylation in Cancer Management

DNA methylation can be exploited on three translational fronts for clinical purposes in cancer patients.

1. *New lines of treatment based on DNA demethylation agents that reverse the CpG island hypermethylation of tumor suppressor genes.* Unlike genetic changes in cancer, epigenetic changes are potentially reversible. For years, in cultured cancer cell lines, we have been able to reexpress genes that had been silenced by methylation by using demethylating agents such as 5-aza-2-deoxycytidine [11]. These compounds had previously been used in the clinic, but the doses administered at those times were quite toxic. Interestingly, we can reduce the doses by adding inhibitors of histone deacetylases, such as phenylbutyrate. Several Phase I and II clinical trials are underway to test this strategy. Chapters 12–14 are excellent guides to understanding this complex area.

2. *Methylation as a molecular biomarker of cancer cells.* The presence of CpG island hypermethylation of the tumor suppressor genes described is specific to transformed cells [5–7] and there is a particular profile of methylation for each tumor type [3,12]. Methylation can, therefore, be used as an indicator for the presence of a particular malignancy. One of the best-accepted cases is the presence of hypermethylation of the glutathione-S-transferase P1 (GSTP1) gene in prostate cancer, as summarized in Chapter 2. Hypermethylation could also be used as a tool for detecting cancer cells in multiple biological fluids [2] or for monitoring hypermethylated promoter loci in serum DNA from cancer patients [13]. Chapters 2 and 5 provide useful information on this matter.

3. *Gene promoter hypermethylation as a prognostic/predictive factor.* Methylation is not only a positive marker, but also a qualitative one. We have recently provided compelling evidence about its strength: the methylation-associated silencing of MGMT (the DNA repair gene) in tumors indicates which patients will be sensitive to chemotherapy with certain alkylating agents [10]. Similar scenarios can now be outlined using the methylation status of hormone and growth factor receptor genes, and those encoding for DNA repair proteins, for many tumor types. The involvement of industry in developing some of these uses in a standardized manner is essential, and it is encouraging that some companies, such as Oncomethylome Sciences, are rising to the challenge.

Global Genomic Hypomethylation

At the same time the CpG islands become hypermethylated, the genome of the cancer cell undergoes global hypomethylation. The malignant cell can have 20 to 60% less genomic 5mC (5-methylcytosine) than its normal counterpart [14,15]. The loss of methyl groups is accomplished mainly by hypomethylation of the "body" (coding region and introns) of genes, and through demethylation of repetitive DNA sequences, which account for 20 to 30% of the human genome. How does global DNA hypomethylation contribute to carcinogenesis? Three mechanisms can be invoked: chromosomal instability, reactivation of transposable elements, and loss of imprinting. Undermethylation of DNA might favor mitotic recombination, leading to loss of heterozygosity, as well as promoting karyotypically detectable rearrangements. Additionally, extensive demethylation in centromeric sequences is common in human tumors and may play a role in aneuploidy. As evidence of this, patients with germline mutations in DNA methyltransferase 3b (DNMT3b) are known to have numerous chromosome aberrations [16]. Hypomethylation of malignant cell DNA can also reactivate intragenomic parasitic DNA, such as L1 (LINE), and Alu (recombinogenic sequence) repeats [17]. These, and other previously silent transposons, may now be transcribed and even "moved" to other genomic regions, where they can disrupt normal cellular genes. The loss of methyl groups can affect imprinted genes and genes from the methylated X chromosome of women. The best-studied case is of the effects of the H19/IGF-2 locus on chromosome 11p15 in certain childhood tumors [18]. Chapter 3 reflects the relevance of DNA methylation to the control of gene imprinting and X inactivation.

In summary, the disruption of DNA methylation patterns is a major hallmark of cancer. Much is still unknown, but the unfolding scenario shows great promise for a better understanding of cancer biology and for improvement in the management of human tumors.

DNA METHYLATION IN IMMUNOLOGY

DNA methylation also occupies a place at the crossroads of many pathways in immunology, providing us with a clearer understanding of the molecular network of the immune system.

From the classical genetic standpoint, two immunodeficiency syndromes, the ICF (immunodeficiency, centromeric regions instability, facial anomalies) syndrome and ATR-X (X-linked form of syndromal retardation associated with alpha-thalassemia) syndrome, are caused by germline mutations in two epigenetic genes: the DNMT3b and the ATRX genes [19,20]. DNMT3b is the putative *de novo* DNA methyltransferase (DNMT1 would be the maintenance DNA methyltransferase and DNMT3a the other *de novo* methyltransferase). In the rare ICF syndrome, characterized by DNA hypomethylation and chromosomal aberration at certain satellite regions, some lymphogenesis genes are expressed in a deregulated fashion [19]. On the other hand, the ATRX gene is a chromatin-remodeling gene that, when mutated, also causes DNA methylation changes, thereby revealing the intimate relationship between chromatin and methylation [20].

Autoimmunity and DNA methylation can also go hand in hand. Classical autoimmune diseases, such as systemic lupus erythematosus or rheumatoid arthritis, are characterized by massive genomic hypomethylation [21,22]. This phenomenon is highly reminiscent of the global demethylation observed in the DNA of cancer cells compared with their normal-tissue counterparts [2,15]. Other examples include the proposed epigenetic control of the histo-blood group ABO genes [23] and the silencing of human leukocyte antigen (HLA) class I antigens [24]. The melanoma antigen gene (MAGE) family of genes provides another illustrative example of immune response mediated by DNA methylation. These genes are not expressed in normal cells because their CpG islands are hypermethylated (contrary to the dogma). However, certain processes, such as malignant transformation, demethylate the island, causing the genes to be reexpressed, with the result that their products are recognized as tumor-specific antigens by cytolytic T lymphocytes [25].

DNA Methylation in Neurosciences, Cardiovascular Research, Metabolic Diseases, Imprinting Disorders, Development, and Cloning

Aberrant DNA methylation patterns go beyond the fields of oncology and immunology to touch a wide range of biomedical and scientific knowledge. In neurology and autism research, for example, it was surprising to discover that germline mutations in the methyl-binding protein MeCP2 (a key element in the silencing of gene expression mediated by DNA methylation) causes the common neurodevelopmental disease known as Rett syndrome [26,27]. This leads us to wonder how many DNA methylation alterations underlie other, more prevalent neurological pathologies, such as schizophrenia or Alzheimer's disease. Beyond that, DNA methylation changes are also known to be involved in cardiovascular disease, the biggest killer in Western countries. For example, aberrant CpG island hypermethylation has been described in atherosclerotic lesions [28]. Germline variants and mutations in genes involved in the metabolism of the methyl group (such as MTHFR) cause changes in DNA methylation [29], and changes in the levels of methyl acceptors and methyl donors are responsible for the pathogenesis of diseases related to homocysteinemia and spina bifida [30]. Imprinting disorders, which represent another huge area of research, are the perfect example of methylation-dependent epigenetic human diseases. A perfectly confined DNA methylation change causes Beckwith-Wiedemann syndrome, Prader-Willi/Angelman syndromes, Russell-Silver syndrome, and Albright hereditary osteodystrophy. This highlights the absolute necessity to maintain the correct DNA methylation pattern in order to achieve harmonized development, as has been demonstrated in mouse models, and is beautifully explained in Chapter 11. Finally, I cannot fail to mention a hot topic: cloning. Perfecting the cloning process will require us to design a system to maintain the fidelity of DNA methylation patterns, whereby we may overcome all the problems that are beginning to come to light in this area in cloned animals [31].

METHODS AVAILABLE FOR THE STUDY OF DNA METHYLATION

Apart from its great value in research into many normal processes and pathological entities, interest in DNA has also boomed as a result of the development of a myriad of powerful and exciting new technologies facilitating its study. The emergence of a new technology for studying DNA methylation, based on bisulfite modification coupled with PCR, has been decisive and is described in detail in

Chapters 4 and 5. It merits a more detailed explanation here. Until a few years ago, the study of DNA methylation was almost entirely based on the use of enzymes that distinguished unmethylated and methylated recognition sites. This approach had many drawbacks, from incomplete restriction cutting to limitation of the regions of study. Furthermore, it usually involved Southern blot technologies, which required relatively substantial amounts of DNA of high molecular weight. The popularization of the bisulfite treatment of DNA (which changes unmethylated "C" to "T" but maintains the methylated "C" as a "C"), associated with amplification by specific polymerase chain reaction primers (methylation-specific polymerase chain reaction), TaqMan, restriction analysis, and genomic sequencing [32], has made it possible for every laboratory and hospital in the world to have a fair opportunity to study DNA methylation, even using pathological material from old archives. We may call this change the "universalization of DNA methylation." The techniques described, which are ideal for studying biological fluids and the detailed DNA methylation patterns of particular tumor suppressor genes, can also be coupled with global genomic approaches for establishing molecular signatures of tumors based on DNA methylation markers, such as CpG island microarrays, restriction landmark genomic scanning (RLGS), and amplification of intermethylated sites (AIMS) [33], as are extensively discussed in Chapters 6–8.

Moreover, we now have serious cause to believe that we can study the content and distribution of 5-methylcytosine in the cellular nuclei and the whole genome, thanks to two new tools: the improved immunohistochemical staining of 5-methylcytosine [34], which allows localization of the latter in the chromatin structure, and high-performance capillary electrophoresis (HPCE), which is a reliable and affordable technique for measuring total levels of 5-methylcytosine [35]. These are described in the practical Chapters 9 and 10.

CONCLUSIONS AND THOUGHTS

The field of DNA methylation is attracting the interest of many researchers and clinicians around the world. Some of the best laboratories are gradually changing their old interests and are moving into the fields of epigenetics and, particularly, DNA methylation. The area is hugely exciting: it combines questions about basic processes (How are DNA methylation patterns established? What key molecules are involved in the mechanism?), and extremely important clinical questions (Is the hypermethylation of this tumor suppressor gene a good marker of poor prognosis, or good response to chemotherapy? Can we use DNA demethylating drugs in chemotherapy regimens?). It also touches on the biology of other more experimental models, such as the mouse, as is skillfully introduced in Chapter 11. Moreover, the influence of DNA methylation is spreading to many disciplines of scientific knowledge: neuroscience, cardiovascular science, genetics, imprinting, sterility, agriculture, immunology, cloning, and so forth (see Figure 1.1).

We sincerely hope, by the multidisciplinary and practical approach we have taken, that this book may serve as an extremely useful starting point for the reader interested in DNA methylation.

> DNA
> METHYLATION

> NORMAL FUNCTIONS:
> CHROMOSOMAL STABILITY, SILENCING OF PARASITIC AND VIRAL SEQUENCES,
> IMPRINTING, X-CHROMOSOME INACTIVATION, TISSUE-SPECIFIC EXPRESSION,
> COMPARTAMENTALIZATION OF CHROMATIN

> ABERRANT DNA METHYLATION:
> METHYLATION-ASSOCIATED SILENCING OF TUMOR SUPPRESSOR,
> GENES AND GLOBAL GENOMIC HYPOMETHYLATION IN CANCER,
> SIMILAR CHANGES OCCUR IN CARDIOVASCULAR AND IMMUNE DISEASE,
> DISRUPTION OF IMPRINTING AND X-CHROMOSOME INACTIVATION,
> ERRONEOUS CLONING

> STUDYING DNA METHYLATION:
> BISULFITE-PCR TECHNOLOGY, HPCE, 5-METHYLCYTOSINE STAINING,
> GLOBAL GENOMIC SCREENINGS (CpG ARRAYS, RLGS, AIMS),
> DNA DEMETHYLATING DRUGS, MOUSE MODELS

FIGURE 1.1 DNA methylation: health, disease, and methodology.

REFERENCES

1. Jenuwein T, Allis CD, Translating the histone code. *Science*, 293, 1074–1080, 2001.
2. Esteller M, Relevance of DNA methylation in the management of cancer. *Lancet Oncol.*, 4, 351–358, 2003.
3. Esteller M, Corn PG, Baylin SB, Herman JG, A gene hypermethylation profile of human cancer. *Cancer Res.*, 61, 3225–3229, 2001.
4. Costello JF, Fruhwald MC, Smiraglia DJ, Rush LJ, Robertson GP, Gao X, Wright FA, Feramisco JD, Peltomaki P, Lang JC, Schuller DE, Yu L, Bloomfield CD, Caligiuri MA, Yates A, Nishikawa R, Su Huang H, Petrelli NJ, Zhang X, O'Dorisio MS, Held WA, Cavenee WK, Plass C, Aberrant CpG-island methylation has non-random and tumour-type-specific patterns. *Nat. Genet.*, 24, 132–138, 2001.
5. Jones, PA, Laird, PW, Cancer epigenetics comes of age. *Nat. Genet.*, 21, 163–167, 1999.
6. Esteller M, CpG island hypermethylation and tumor suppressor genes: a booming present, a brighter future. *Oncogene*, 21, 5427–5440, 2002.
7. Herman JG, Baylin, SB, Gene silencing in cancer in association with promoter hypermethylation. *N. Engl. J. Med.*, 349, 2042–2054, 2003.
8. Esteller M, Garcia-Foncillas J, Andion E, Goodman SN, Hidalgo OF, Vanaclocha V, Baylin SB, Herman JG, Inactivation of the DNA-repair gene MGMT and the clinical response of gliomas to alkylating agents. *N. Engl. J. Med.*, 343, 1350–1354, 2000.
9. Esteller M, Gaidano G, Goodman SN, Zagonel V, Capello D, Botto B, Rossi D, Gloghini A, Vitolo U, Carbone A, Baylin SB, Herman JG, Hypermethylation of the DNA repair gene O(6)-methylguanine DNA methyltransferase and survival of patients with diffuse large B-cell lymphoma. *J. Natl. Cancer Inst.*, 94, 26–32, 2002.

10. Esteller M, Generating mutations but providing chemosensitivity: the role of O6-methylguanine DNA methyltransferase in human cancer. *Oncogene*, 23, 1–8, 2004.

11. Villar-Garea A, Esteller M, DNA demethylating agents and chromatin-remodelling drugs: which, how and why? *Curr. Drug Metab.*, 4, 11–31, 2003.

12. Paz MF, Fraga MF, Avila S, Guo M, Pollan M, Herman JG, Esteller M, A systematic profile of DNA methylation in human cancer cell lines. *Cancer Res.*, 63, 1114–1121, 2003.

13. Esteller M, Sanchez-Cespedes M, Rosell R, Sidransky D, Baylin SB, Herman JG, Detection of aberrant promoter hypermethylation of tumor suppressor genes in serum DNA from non-small-cell lung cancer patients. *Cancer Res.*, 59, 67–70, 1999.

14. Ehrlich M, DNA hypomethylation and cancer. In: *DNA alterations in cancer: genetic and epigenetic changes,* ed. Melanie Ehrlich, Eaton Publishing, Natick, MA, 273–291, 2000.

15. Esteller M, Fraga MF, Guo M, Garcia-Foncillas J, Hedenfalk I, Godwin AK, Trojan J, Vaurs-Barriere C, Bignon YJ, Ramus S, Benitez J, Caldes T, Akiyama Y, Yuasa Y, Launonen V, Canal MJ, Rodriguez R, Capella G, Peinado MA, Borg A, Aaltonen LA, Ponder BA, Baylin SB, Herman JG, DNA methylation patterns in hereditary human cancers mimic sporadic tumorigenesis. *Hum. Mol. Genet.*, 10, 3001–3007, 2001.

16. Xu GL, Bestor TH, Bourc'his D, et al., Chromosome instability and immunodeficiency syndrome caused by mutations in a DNA methyltransferase gene. *Nature*, 1999; 402. 187–191, 1999.

17. Yoder JA, Walsh CP, Bestor TH, Cytosine methylation and the ecology of intragenomic parasites. *Trends Genet.*, 13, 335–340, 1997.

18. Plass C, Soloway PD, DNA methylation, imprinting and cancer. *Eur. J. Hum. Genet.*, 10, 6–16, 2002.

19. Okano M, Bell DW, Haber DA, Li E, DNA methyltransferases Dnmt3a and Dnmt3b are essential for de novo methylation and mammalian development. *Cell.*, 99, 247–257, 1999.

20. Gibbons RJ, McDowell TL, Raman S, O'Rourke DM, Garrick D, Ayyub H, Higgs DR, Mutations in ATRX, encoding a SWI/SNF-like protein, cause diverse changes in the pattern of DNA methylation. *Nat. Genet.*, 24, 368–371, 2000.

21. Richardson B, Scheinbart L, Strahler J, Gross L, Hanash S, Johnson M, Evidence for impaired T cell DNA methylation in systemic lupus erythematosus and rheumatoid arthritis. *Arthritis Rheum.*, 33, 1665–1673, 1990.

22. Corvetta A, Della Bitta R, Luchetti MM, Pomponio G, 5-Methylcytosine content of DNA in blood, synovial mononuclear cells and synovial tissue from patients affected by autoimmune rheumatic diseases. *J. Chromatogr.* 566, 481–491, 1991.

23. Kominato Y, Hata Y, Takizawa H, Tsuchiya T, Tsukada J, Yamamoto F, Expression of human histo-blood group ABO genes is dependent upon DNA methylation of the promoter region. *J. Biol. Chem.*, 274, 37240–37250, 1999.

24. Nie Y, Yang G, Song Y, Zhao X, So C, Liao J, Wang LD, Yang CS, DNA hypermethylation is a mechanism for loss of expression of the HLA class I genes in human esophageal squamous cell carcinomas. *Carcinogenesis*, 22, 1615–1623, 2001.

25. De Smet C, Lurquin C, Lethe B, Martelange V, Boon T, DNA methylation is the primary silencing mechanism for a set of germ line- and tumor-specific genes with a CpG-rich promoter. *Mol. Cell Biol.*, 19, 7327–7335, 1999.

26. Shahbazian MD, Zoghbi HY, Rett syndrome and MeCP2: linking epigenetics and neuronal function. *Am. J. Hum. Genet.*, 71, 1259–1272, 2002.

27. Ballestar E, Paz MF, Valle L, Wei S, Fraga MF, Espada J, Cigudosa JC, Huang TH, Esteller M, Methyl-CpG binding proteins identify novel sites of epigenetic inactivation in human cancer. *EMBO J.,* 22, 6335–6345, 2003.

28. Post WS, Goldschmidt-Clermont PJ, Wilhide CC, Heldman AW, Sussman MS, Ouyang P, Milliken EE, Issa JP, Methylation of the estrogen receptor gene is associated with aging and atherosclerosis in the cardiovascular system. *Cardiovasc Res.,* 43, 985–991, 1999.

29. Paz MF, Avila S, Fraga MF, Pollan M, Capella G, Peinado MA, Sanchez-Cespedes M, Herman JG, Esteller M, Germ-line variants in methyl-group metabolism genes and susceptibility to DNA methylation in normal tissues and human primary tumors. *Cancer Res.,* 62, 4519–4524, 2002.

30. Eskes TK, Neural tube defects, vitamins and homocysteine. *Eur. J. Pediatr.,* 157, Suppl. 2: S139–141, 1998.

31. Shi W, Zakhartchenko V, Wolf E, Epigenetic reprogramming in mammalian nuclear transfer. *Differentiation,* 71, 91–113, 2003.

32. Fraga MF, Esteller M, DNA methylation: a profile of methods and applications. *Biotechniques,* 33, 632–649, 2002.

33. Paz MF, Wei S, Cigudosa JC, Rodriguez-Perales S, Peinado MA, Huang TH, Esteller M, Genetic unmasking of epigenetically silenced tumor suppressor genes in colon cancer cells deficient in DNA methyltransferases. *Hum. Mol. Genet.,* 12, 2209–2219, 2003.

34. Piyathilake CJ, Johanning GL, Frost AR, Whiteside MA, Manne U, Grizzle WE, Heimburger DC, Niveleau A, Immunohistochemical evaluation of global DNA methylation: comparison with in vitro radiolabeled methyl incorporation assay. *Biotech. Histochem.,* 75, 251–258, 2000.

35. Fraga MF, Uriol E, Borja Diego L, Berdasco M, Esteller M, Canal MJ, Rodriguez R, High-performance capillary electrophoretic method for the quantification of 5-methyl 2′-deoxycytidine in genomic DNA: application to plant, animal and human cancer tissues. *Electrophoresis,* 23, 1677–1681, 2002.

2 Uses of DNA Methylation in Cancer Diagnosis and Risk Assessment

Carmen Jerónimo, Rui Henrique, and David Sidransky

CONTENTS

ACKNOWLEDGMENTS

C.J. and R.H. are recipients of grants from Fundação para a Ciência e Tecnologia, (SFRH/BPD 8031/2002) and Liga Portuguesa Contra o Cancro — Núcleo Regional do Norte, Portugal, respectively. Supported by an NCI grant from the Early Detection Research Network and a collaborative agreement between Oncomethylome Sciences and D.S.

INTRODUCTION

Over the last decade, molecular markers have emerged as promising tools for early cancer detection, patient management, and assessment of prognosis. Research into nucleic acid-based markers has blossomed as their ability to allow for robust analysis of clinical samples (tissue or body fluids) in a high-throughput fashion [1] has become recognized. Among the DNA-based approaches, DNA methylation offers several advantages over other types of molecular markers. Promoter hypermethylation of key regulatory genes in human cancers is a frequent event and is often associated with transcription silencing [2]. In addition, most of the common human tumors appear to have one or more hypermethylated loci when several methylation markers are examined together [3]. Furthermore, a specific pattern of hypermethylation for each type of human cancer is emerging from the rapidly growing list of cancer-related methylated genes, eventually permitting the identification of the tissue of origin of a particular neoplasm [3]. Notably, promoter methylation occurs early in tumorigenesis and may be identified even in preneoplastic lesions [4–6]. From a methodological standpoint, the identification of methylated DNA is easier than RNA manipulation followed by reverse transcription-PCR, since DNA harboring methylation is more stable. Likewise, detection of methylated DNA is particularly suited for clinical samples because methylation is a positive signal that is less likely to be masked through contaminant normal DNA. Thus, methylation markers provide sensitive detection in samples where tumor DNA is scarce or diluted by excess normal DNA.

To be used in the clinical setting, methylation marker technologies are required to possess high sensitivity and specificity, reproducibility, homogeneity, and high-throughtput capabilities. The methylation-specific PCR (MSP) assay [7] allows for the fast and precise identification of methylated DNA at gene promoter regions that correlates with transcriptional abrogation. In the MSP assay, DNA is treated with bisulfite, converting unmethylated cytosines to uracil, whereas methylated cytosines are protected and remain unchanged. Due to its relative simplicity, safety (it does not use radioactivity), and sensitivity, MSP is a clinically useful method for promoter hypermethylation detection. Thus far, this method has permitted the identification of a large number of methylated genes in several human cancers. However, qualitative results provided by MSP may mask relevant quantitative differences.

The development of a specific fluorescence-based real-time quantitative MSP (QMSP) assay in recent years has represented an important step forward in the acquisition of information on DNA methylation obtained from clinical specimens [8]. The continuous monitoring of the fluorescent signals during the PCR process enables quantification of methylated alleles of a single region among unmethylated DNA, because the fluorescence emission represents the number of generated DNA fragments [9]. The sensitivity of QMSP is comparable to conventional MSP (1:1,000 to 1:10,000), but the specificity is higher due to the more stringent amplification conditions and the use of the oligonucleotide probe. Likewise, QMSP allows for the rapid analysis of multiple markers in a large number of samples and does not require the use of electrophoretic gels, thus minimizing contamination of subsequent reactions due to manipulation of PCR products. Indeed, methodologies that enable the analysis of multiple molecular markers are likely to provide more relevant clinical information than single marker analysis. Moreover, these assays are amenable to large-scale screening of several target genes in multiple samples.

Noninvasive molecular detection of cancer requires the use of highly sensitive assays that may detect minimal amounts of altered DNA obtained from body fluids of cancer patients. Studies have suggested for over two decades that the amount of DNA in serum of cancer patients is higher than in normal subjects [10], thus providing a valuable means for screening several cancer types in a single sample. Besides DNA extracted from serum and plasma [11–15], other types of body fluids and clinical samples have been successfully tested for hypermethylation of cancer-related genes, including urine and semen [16–18], bronchoalveolar lavage fluid (BAL) [19], saliva [20], sputum [21,22], ductal lavage fluid [23], and fine-needle aspirates [24]. As highlighted below, high-throughput automation of these assays might provide clinically valuable information, ranging from early diagnosis and risk assessment, to therapeutic decision making, prognosis, and surveillance.

HEAD AND NECK CANCER

Head and neck cancer (HNC) accounts for an important proportion of cancer-related morbidity and mortality [25]. Local and regional recurrences are frequent events that impair continuing efforts to improve patient outcome. Thus, the role of early detection using methylation markers has been emphasized. Several gene promoters have been reported to be frequently and specifically hypermethylated in HNC patients, namely *p16, p15, MGMT*, and *DAPK* [12,26–28]. Remarkably, methylation at these loci was found in up to 55% of serum and 56% of saliva DNA samples collected from HNC patients [12,20]. In nasopharyngeal carcinoma, a common type of HNC in Asian countries, gene promoter methylation at several loci was successfully detected in nasopharyngeal swabs, throat-rinsing fluids, plasma, and peripheral blood [29–31]. Hence, these markers might be used as molecular tools for early cancer detection in at-risk populations and also for patient surveillance. This approach is further supported by the fact that *p16* promoter methylation is already present in preneoplastic lesions, such as oral epithelial dysplasia [32]. Although *p16* and *p15* methylation in plasma DNA from normal subjects was recently reported (20% and 50%, respectively), methylation levels were significantly lower when compared to HNC patients, once again emphasizing the need for quantitative MSP assays [27]. In most studies, no significant associations were found between tumor methylation status and clinical pathological parameters, though a significant correlation between tumor invasiveness and *DCC* methylation was reported [26]. In the same study, methylation of *MINT31* emerged as an independent predictor of outcome [26], but these findings need confirmation in a larger series of patients.

LUNG CANCER

Lung cancer is the most frequent cause of cancer-associated mortality in the United States and European countries [33]. Early diagnosis strategies are urgently needed because two thirds of patients present an advanced (i.e., noncurable) disease stage at initial clinical diagnosis [34]. Lung cancer comprises a number of distinct pathological entities that for practical purposes are grouped in two major types: small-cell lung carcinoma (SCLC) and non-small-cell lung carcinoma (NSCLC). The latter group is the more frequent (~80%) and the most extensively studied from an epigenetic standpoint, but is also the more heterogeneous. Among the several epigenetic markers tested in lung cancer patients, promoter methylation of *p16, MGMT, DAPK, RASSF1A*, and *APC* showed promising results. *P16* hypermethylation was detected in 100%, 63%, and 33% of sputum, BAL, and serum DNA samples, respectively [11,19,22], whereas *MGMT* methylation was found in 66% of sputum and 66% of serum DNA samples from lung cancer patients [11,22]. Remarkably, hypermethylation at these 2 gene promoters was detected 5 to 35 months prior to clinical diagnosis [22]. These results are further supported by recent reports of aberrant promoter methylation of multiple genes in current and former smokers [35,36]. Analysis of methylation at selected genes in sputum DNA from individuals at high risk for developing lung cancer thus represents a promising early cancer-detection strategy. Methylation assays were shown to be more sensitive than conventional cytology of sputum or BAL for lung cancer detection [22,37]. The specificity of some PCR assays may be impaired by the reported detection of gene methylation in normal lung tissue and even normal lymphocytes from healthy individuals [22,37–39].

The *APC* promoter 1A is also frequently hypermethylated in lung cancer (95% of tumors) and is detectable in approximately 50% of serum DNA samples from NSCLC patients, irrespective of clinical stage [15,38]. Although methylation at this locus was identified in normal lung tissue, a QMSP assay demonstrated that methylation levels were significantly higher in tumors [38]. Lower *APC* and *MGMT* methylation levels were shown to be associated with improved survival [15,38,40]. In multivariate analysis, quantitative *APC* and *MGMT* methylation were found to be independent prognostic factors, and similar results were reported for *DAPK* and *RASSF1A* methylation in stage I NSCLC patients, using conventional MSP [41,42]. Furthermore, QMSP assays might also prove useful for occult metastasis detection, providing a more accurate staging approach for resectable NSCLC [43].

DIGESTIVE TRACT CANCERS

Several studies suggest that aberrant promoter methylation is an early event in gastrointestinal (GI) carcinogenesis. Indeed, hypermethylation of genes with a critical function in cell cycle regulation, such as *p16*, *p14*, and *APC*, was reported in colorectal adenomas [5,44] and even normal gastric mucosa adjacent to tumor [45]. Moreover, aging-associated methylation of several genes (e.g., *ESR*, *MyoD1*, *IGF2*, *N33*, *PAX6*, and *CSPG2*) was observed in normal colonic mucosa and might be related to the increased risk for GI cancer in the elderly [46–49]. It is hypothesized that age-related methylation might contribute to the development of a field defect enabling the initiation and progression of cancer in older subjects, although the individual gene susceptibility to methylation seems to vary considerably and is likely to be tissue-specific. In contrast, some patterns of gene methylation seem to be more cancer-specific [50]. The clinical significance is illustrated in two studies where the methylation of these genes was found to be associated with worse prognosis for all cancer stages [51] and even to be an independent predictor of survival in stage III colorectal cancer patients treated with 5-fluorouracil [52]. Detection of DNA methylation in serum as a diagnostic tool for colorectal cancer is also feasible. Methylation of *DAPK*, *hMLH1*, and *p16* gene promoters was found in serum of 21%, 33%, and 30% colorectal cancer patients, respectively [14,53,54]. Interestingly, *p16* methylation in primary tumors was associated with poor prognosis [55], whereas *p16* methylation in regional lymph nodes and in serum might be a useful marker for early detection of tumor recurrence [56,57].

In gastric cancer patients, promoter hypermethylation of one or more of five genes tested (*p15*, *p16*, *DAPK*, *GSTP1*, and *CDH1*) was detected in the serum of 83.3% of patients, but not in controls, and was independent of tumor stage [58]. Nonetheless, in a different study, *p16* methylation alone was detected in only 26% of serum samples from patients with gastric cancers harboring the same epigenetic alteration. These results emphasize the need for analysis of gene panels instead of single genes to increase the sensitivity of molecular detection assays. Yet, due to the high prevalence of gastric cancer and the lack of universally accepted screening tools, detection of specific methylated genes in serum should be further pursued. A correlation between *MGMT* methylation and features of poor prognosis in gastric cancer was reported, but this epigenetic alteration did not bear independent predictive value [59].

APC promoter methylation is the most frequent epigenetic alteration reported to date in esophageal adenocarcinoma, ranging from 68 to 92% of tumors [60,61]. Although no prognostic value has been ascribed to *APC* methylation status in primary tumors, high plasma levels of methylated gene were found to be associated with decreased survival in esophageal adenocarcinoma patients using QMSP [13]. Thus, quantitative *APC* methylation in plasma stands as a promising prognostic biomarker. Using conventional MSP, Brock and coworkers also did not find a correlation between single gene methylation in tumor tissue and clinical pathological parameters [61]. However, when the panel of target genes was analyzed, those tumors with a higher number of methylated loci were significantly associated with poor survival and tumor recurrence [61]. Notably, selected gene panels have the potential to identify premalignant lesions (Barrett's esophagus) with the increased likelihood of progression to adenocarcinoma, with performance that is superior to histology [60]. In esophageal squamous cell carcinoma, the frequency of *APC* methylation in tumor tissue and plasma was considerably lower than in adenocarcinoma (50% and 6.3%, respectively) [13]. In addition, *p16* methylation was present in 82% of primary esophageal tumors and detectable in 23% of sera DNA obtained from the same patients [62].

Aberrant promoter methylation at the *p16* locus is also a common event in hepatocellular carcinoma (67 to 73%) and detection of this epigenetic alteration in serum and plasma was successfully accomplished in over 70% of patients with perfect specificity [63,64]. The use of quantitative assays might improve the detection rate and provide an additional tool for patient monitoring [65]. Moreover, gene methylation profiling studies have suggested the possibility of discriminating cancerous lesions from nonmalignant liver tissue [66]. This analysis could be useful in histological assessment where difficulties often arise in the diagnosis of hepatocellular carcinoma from liver biopsies. In addition, it would be interesting to seek other early epigenetic markers that might aid in the detection of hepatocellular carcinoma in its early stages, thereby improving the efficacy of curative therapeutic options.

Pancreatic cancer is also a major challenge for early detection. Imaging techniques and cytologic examination are not sufficiently sensitive, and genetic testing has not provided specific biomarkers that might differentiate benign tissue from malignant lesions [67]. Although several gene promoters were found to be methylated in pancreatic carcinoma [68,69], only *ppENK* and *p16* were tested as potential biomarkers [70]. In the latter study, *ppENK* was detected in 57% of pancreatic juice samples of patients with periampullary cancer, whereas the detection rate for *p16* was lower (14%). The main drawback of this study was the detection of *ppENK* and *p16* in a significant proportion of duodenal contents from normal subjects although it was absent in normal ductal pancreatic epithelium [70]. Eventually, the use of cannulation procedures that might yield pure pancreatic juice samples might increase the specificity of the test [70].

Breast Cancer

Several genes involved in most cancer-related pathways are known to be methylated in breast cancer [71], and the list is increasing as DNA methylation becomes an appealing target for breast cancer detection. Moreover, promoter methylation is an early and frequent event in breast carcinogenesis [72]. *APC*, *cyclin D2*, *stratifin*, *p16*, and *RASSF1A* are among the more frequently methylated genes in breast cancer [72–77], and several of them were tested as biomarkers. *P16* methylation was detected in 8 to 14% of plasma DNA samples from breast cancer patients, while 11 to 23% of primary tissues showed this epigenetic alteration [78–80]. Aberrant *CDH1* methylation was also detected in 20% of plasma samples from breast cancer patients [80]. Promising results regarding *APC* and *RASSF1A* methylation in serum of breast cancer patients were recently reported [81]. These two markers, either individually or in combination, came out as independent prognostic markers for overall survival after screening from a list of 39 methylated gene targets. Importantly, methylated *APC* was not found in normal controls, and only one control harbored *RASSF1A* methylation [81]. Besides serum DNA, cells collected from the mammary ducts were tested for DNA methylation [23]. *Cyclin D2*, *RAR-*, and *Twist* methylation were detected with a high frequency in ductal fluid collected from cancer patients, while these alterations were seldom observed in fluid from normal ducts. These results might augment cytological assessment of ductal lavage fluid and eventually nipple discharge samples, which might be limited by poor preservation and inadequate representation of the malignant lesion. Gene promoter methylation of breast fine-needle aspirates is also feasible and might be useful as an adjunct to cytological examination [24,82].

Genitourinary Tract Cancer

Among urological malignancies, prostate cancer stands as a leading cause of cancer-related mortality and morbidity in the Western world. The importance of early detection has been emphasized because curative treatment (radical prostatectomy or radiotherapy) is achievable only in the earliest stage of the disease [83]. Because current screening methodologies display low specificity [84], new approaches that might detect prostate cancer at an early stage and accurately discriminate malignancies from benign lesions are clearly needed. Furthermore, prostate cancer diagnosis in biopsies meets with significant difficulties, ranging from inadequate sampling to considerable interobserver variability in diagnostic assessment and grading [85]. Eventually, the emergence of new molecular markers might increase the consistency of prostate cancer diagnosis in sextant needle biopsies and provide more reliable guidelines to define prognosis and therapeutic modalities.

A lack of sufficiently frequent or early genetic alterations that occur in prostate cancer prompted a search for epigenetic alterations as tumor markers. The epigenetic silencing of the glutathione-S-transferase P1 (*GSTP1*) gene is the most common somatic genome alteration reported in prostate cancer, occurring in over 90% of cancers and even in up to 70% of high-grade prostatic intraepithelial neoplasia (HGPIN), a putative prostate cancer precursor lesion [6,86–88]. This gene codes for an enzyme involved in DNA protection from electrophilic metabolites of carcinogens and reactive oxygen species by conjugating chemically reactive electrophiles to glutathione [89]. In prostate cancer, loss of expression of this enzyme is a frequent finding and *GSTP1* gene silencing is directly associated with promoter hypermethylation [86,88,90]. However, since induced *GSTP1* expression in prostate cancer cell lines fails to suppress cell growth [91], *GSTP1* is not recognized as a tumor suppressor gene at the present time and was instead proposed to act as a "caretaker" gene [92].

GSTP1 hypermethylation is detectable in most body fluids of prostate cancer patients, with variable sensitivity and specificity, depending on the methodology and sampling procedure as depicted in Table 2.1. In plasma or serum DNA, conventional MSP performed better than QMSP (up to 72% sensitivity in the former and 13% in the latter), and both demonstrated perfect specificity [16,18]. Although the reasons for this apparent flaw of QMSP are not immediately clear, the superior stringency of the QMSP assay (owing to the use of an internal probe) and the high background level of fluorescence intrinsic to QMSP analysis might be responsible for that result. Further improvements in the QMSP assay and future studies aimed at detecting *GSTP1* methylation in body fluids of prostate cancer patients are expected to improve

TABLE 2.1
Performance of *GSTP1* Promoter Methylation Assays in Prostate Tissue and Body Fluids from Prostate Cancer Patients

	Assay	Sensitivity	Specificity	Reference
Prostate tissue				
Radical prostatectomy	QMSP	79–85%	96.8–100%	[6,99]
Biopsies	QMSP	73–90.9%	100%	[6,98,99]
Plasma/serum	QMSP	13%	100%	[18]
	MSP	36.2–72%	100%	[6,16]
Urine				
Voided urine	QMSP	18.8%	93%	[18]
	MSP	30–36%	95–100%	[16,18]
After prostatic massage	MSP	73%	98%	[93]
Ejaculate	MSP	50%	100%	[16]

the detection rate. Of interest, the study reporting the higher rate of *GSTP1* methylation detection in plasma and serum included a large proportion of patients with advanced disease stage (approximately 45%) not amenable to curative treatment [16]. Thus, the likelihood of these patients having circulating tumor cells was elevated and might be responsible for the high detection rate in plasma and serum samples.

Urine samples constitute an obvious target for urological cancer detection. In prostate cancer patients, *GSTP1* methylation was identified in 27 to 36% voided urine samples using conventional MSP [16–18], whereas QMSP reached only 18.8% sensitivity [18]. The use of prostatic massage prior to urine collection resulted in a higher sensitivity (73%) due to the increased shedding of neoplastic cells in the prostatic ducts [93]. Nevertheless, the specificity of the assay seemed to be compromised because 1 out of 45 patients with benign prostatic hyperplasia (BPH) and 2 out of 7 patients with HGPIN in prostate biopsies also tested positive in the respective urine samples [93]. Although BPH lesions are reported to harbor *GSTP1* methylation, the levels are significantly lower when compared to prostate cancer [6]. Moreover, *GSTP1* methylation was also found in the urine of another BPH patient but no methylation was detected in the respective tissue [18]. Because over 75% of prostate carcinomas arise in the peripheral zone while BPH is a transition zone lesion, and patients with the isolated finding of HGPIN in prostate biopsy have a high probability of being diagnosed with prostate cancer in a subsequent biopsy, it is tempting to speculate whether those patients with BPH or HGPIN simultaneously harbored a prostate carcinoma that would be responsible for the positive *GSTP1* methylation test in urine.

It has also been reported that *GSTP1* methylation analysis performed in urine samples collected after prostate biopsy yielded a higher detection rate, and might prove useful for the stratification of patients into low- and high-risk categories for further clinical investigation [94]. An additional finding of this study was that a significant proportion of men with negative or suspicious prostate biopsy tested positive for *GSTP1* methylation in urine. Since no information about the methylation status of the respective tissue samples was provided, it is difficult to determine if such cases are false positives or are indeed part of the significant number of under-diagnosed cancer cases after prostate biopsy [95,96].

Interestingly, a high detection rate of prostate cancer in ejaculates using conventional MSP for *GSTP1* hypermethylation was reported [16,97]. However, those studies were performed in a rather small set of patients (8 or 9) and the nature of the sampling procedure, particularly in older men, seems to preclude its widespread clinical use.

The role of *GSTP1* promoter methylation for the detection of prostate carcinoma is not limited to body fluids, and the usefulness of this prostate cancer marker is now firmly established in tissue samples. Whereas *GSTP1* methylation was detectable in 29% of BPH and 91.3% of prostate cancer tissues, methylation levels were significantly lower in the former [6]. Thus, a QMSP assay for *GSTP1* could reliably discriminate cancer from noncancerous prostate lesions. Figure 2.1 illustrates typical *GSTP1* amplification plots for prostate carcinoma, HGPIN, and BPH. The difference in methylation levels enabled the setting of a cut-off value, resulting in a detection test with 85.5% sensitivity and 96.8% specificity. When the same QMSP assay was performed blindly in a set of 21 prostate biopsies (10 biopsies

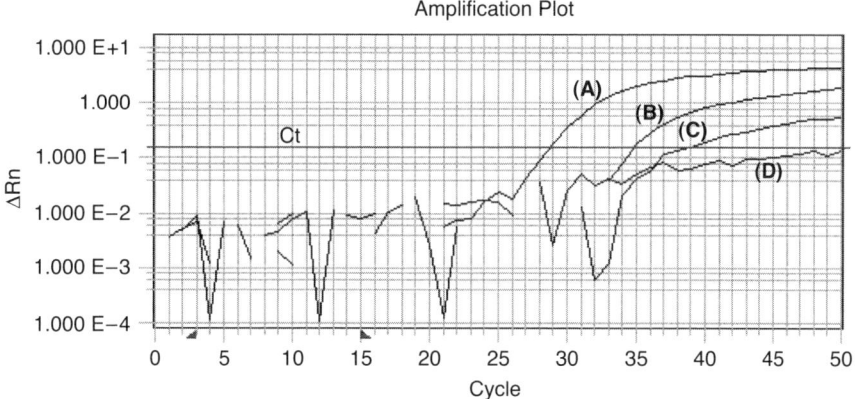

FIGURE 2.1 Illustrative QMSP amplification plots for *GSTP1* from prostate carcinomas (A and D), HGPIN (B), and BPH (C) tissues. The prostate carcinoma depicted in (A) shows stronger amplification of the target gene compared to HGPIN and BPH. The prostate carcinoma corresponding to (D) is unmethylated at the *GSTP1* promoter (the amplification plot does not cross the threshold line, Ct). The *GSTP1/ACTB* ratios determined using the cycle number where fluorescence per reaction crossed the threshold (Ct), which is set to the geometrical phase of polymerase chain reaction amplification above background. ΔRn is defined as the cycle-to-cycle change in the reporter fluorescence signal normalized to a passive reference fluorescence signal (log scale).

without evidence of cancer and 11 with histologically proven adenocarcinoma) *GSTP1* methylation levels correctly predicted the histologic diagnosis with 90.9% sensitivity, 100% specificity, and 100% positive predictive value [6]. Subsequently, quantitative *GSTP1* methylation was able detect the presence of minute foci of prostate cancer in sextant prostate biopsies and accurately discriminate malignant from benign prostate lesions in limited tissue samples [98]. In a recent prospective study, the QMSP assay for *GSTP1* alone showed higher sensitivity than histopathology alone performed by an expert uropathologist (75% vs. 64%) [99]. Furthermore, the combination of histopathology and QMSP for *GSTP1* reached a sensitivity of 79%, with 100% specificity, representing a 15% increase over histopathology alone [99]. It is of note that these results were attained with 100% specificity. Perfect specificity is a key issue in prostate cancer detection due to the therapeutic implications of a diagnosis of malignancy.

Some concern might arise regarding the putative clinical relevance of prostate tumors detected by molecular means. Indeed, a variable proportion of prostate carcinomas seem to be clinically silent and apparently would not be detected if screening tests were not available [100]. Correlations between the methylation index (i.e., the proportion of methylated genes in a given panel) and markers of poor outcome in prostate cancer have been reported [101]. Hence, future studies should determine whether carcinomas detected in prostate biopsy due to high levels of promoter methylation are more likely to benefit from appropriate therapeutic intervention.

Although *GSTP1* hypermethylation was reported in breast and liver cancer [3,102], *GSTP1* hypermethylation is considerably less frequent in other genitourinary malignancies, namely, bladder and renal cancer [3,103–105], increasing the specificity of this marker for prostate cancer detection. In bladder cancer, frequent methylation was reported for *RAR-*, *DAPK*, *APC*, *RASSF1A*, *CDH1*, *CDH13*, and *p16* [104,105]. Whereas bladder cancer patients harboring tumors with high methylation index seem to have a shorter survival [104], *CDH1* methylation-positive status emerged as an independent prognostic parameter [104], and DAPK hypermethylation was found to be associated with early recurrence of superficial disease [106]. Of interest, the methylation analysis of a panel of four genes (*RAR-*, *DAPK*, *CDH1*, and *p16*) in voided urine was able to yield a higher bladder cancer detection rate than conventional urine cytology (90.9% vs. 45.5%), most strikingly in low-grade cancers [105]. Nonetheless, the specificity of the assay was only 76.4% due to the detection of *RAR-* in urine from controls. The addition of *RASSF1A* to this panel might increase the power of the assay because *RASSF1A* methylation was detected in 50% voided urine from bladder cancer patients, whereas it was completely negative in controls [107]. Likewise, methylation analysis of plasma DNA for bladder cancer detection was attempted. The methylation status of *p16* and *p14* were assessed, but the sensitivity was less than 50% for *p16*, though the specificity reached 95% [108,109]. In addition, *p14* methylation in plasma was associated with tumor size, multicentricity, and relapse [108].

Information regarding the gene methylation profile of renal cell carcinoma is relatively scarce [3,110,111]. *DAPK*, *MT1G*, *NORE1A*, and *CDH1* are methylated at low frequency (<25%), whereas *RASSF1A* and *TIMP-3* are methylated in over 50% of cases. Like prostate and bladder cancer, renal cell cancer is amenable to detection in urine and a survey for methylation of selected tumor suppressor genes was recently reported [111]. A conventional MSP assay detected an identical pattern of gene hypermethylation in 88% of matched urine DNAs, including most patients with localized disease. Because no hypermethylation was detected for any of the panel genes in the controls, which included patients with benign kidney disease, the test showed perfect specificity [111]. Larger studies, which simultaneously include patients with different urological cancers and confounding genitourinary diseases, will shed light on the ultimate use of urine DNA for diagnosis and monitoring.

METHYLATION AND PREDICTION OF RESPONSE TO CHEMOTHERAPY

DNA methylation-based markers also provide relevant information regarding response to chemotherapy in some tumors. The most illustrative example is the association between *MGMT* silencing via promoter methylation and tumor sensitivity to chemotherapy with carmustine in brain gliomas [112]. Due to hypermethylation, the DNA repair function of *MGMT* is impaired, and neoplastic cells become more susceptible to damage induced by alkylating agents. Increased response to chemotherapy with cyclophosphamide and improved prognosis were also reported for patients with diffuse large B-cell lymphoma harboring *MGMT* promoter methylation

[113]. An additional therapeutical application of methylation analysis results from the reported association between *FANCF* promoter methylation (which renders cells more susceptible to DNA damaging agents) and chemosensitivity of ovarian cancer cells to cisplatin [114]. Further dissection of the epigenome in several tumor models is likely to unravel more clinically useful markers.

FUTURE PROSPECTS

The emergence of DNA methylation markers holds the promise of accurate, sensitive, and cost-effective cancer detection in body fluids or tissue samples obtained by noninvasive or minimally invasive techniques. Panels of markers that include the most frequently epigenetically altered genes might provide important tools for risk assessment, early diagnosis, therapeutic decision making, disease monitoring, and prognosis. These panels will likely include quantitative assessment of markers such as *GSTP1*, *APC*, *MGMT*, *p16*, *DAPK*, and *RASSF1A*, which either tested individually, or together, may be able to cover the more common human cancers. We are just starting to unravel the vast clinical potential of methylation markers and, in the process, transforming the field of epigenetics into one of the most exciting areas in translational cancer research.

REFERENCES

1. Sidransky, D., Emerging molecular markers of cancer, *Nat. Rev. Cancer*, 2, 210, 2002.
2. Herman, J.G. and Baylin, S.B., Gene silencing in cancer in association with promoter hypermethylation, *N. Engl. J. Med.*, 349, 2042, 2003.
3. Esteller, M. et al., A gene hypermethylation profile of human cancer, *Cancer Res.*, 61, 3225, 2001.
4. Esteller, M. et al., hMLH1 promoter hypermethylation is an early event in human endometrial tumorigenesis, *Am. J. Pathol.*, 155, 1767, 1999.
5. Esteller, M. et al., Analysis of adenomatous polyposis coli promoter hypermethylation in human cancer. *Cancer Res.*, 60, 4366, 2000.
6. Jerónimo, C. et al., Quantitation of GSTP1 methylation in non-neoplastic prostatic tissue and organ-confined prostate adenocarcinoma, *J. Natl. Cancer Inst.*, 93, 1747, 2001.
7. Herman, J.G. et al., Methylation-specific PCR: a novel PCR assay for methylation status of CpG islands, *Proc. Natl. Acad. Sci. USA*, 93, 9821, 1996.
8. Eads, C.A. et al., MethyLight: a high-throughput assay to measure DNA methylation. *Nucleic Acids Res.*, 28, E32, 2000.
9. Heid, C.A. et al., Real time quantitative PCR, *Genome Res.*, 6, 986, 1996.
10. Shapiro, B. et al., Determination of circulating DNA levels in patients with benign or malignant gastrointestinal disease, *Cancer*, 51, 2116, 1983.
11. Esteller, M. et al., Detection of aberrant promoter hypermethylation of tumor suppressor genes in serum DNA from non-small cell lung cancer patients, *Cancer Res.*, 59, 67, 1999.
12. Sanchez-Cespedes, M. et al., Gene promoter hypermethylation in tumors and serum of head and neck cancer patients, *Cancer Res.*, 60, 892, 2000.

13. Kawakami, K. et al., Hypermethylated APC DNA in plasma and prognosis of patients with esophageal adenocarcinoma, *J. Natl. Cancer Inst.*, 92, 1805, 2000.

14. Grady, W.M. et al., Detection of aberrantly methylated hMLH1 promoter DNA in the serum of patients with microsatellite unstable colon cancer, *Cancer Res.*, 61, 900, 2001.

15. Usadel, H. et al,. Quantitative adenomatous polyposis coli promoter methylation analysis in tumor tissue, serum, and plasma DNA of patients with lung cancer, *Cancer Res.*, 62, 371, 2002.

16. Goessl, C. et al., Fluorescent methylation-specific polymerase chain reaction for DNA-based detection of prostate cancer in bodily fluids, *Cancer Res.*, 60, 5941, 2000.

17. Cairns, P. et al., Molecular detection of prostate cancer in urine by GSTP1 hyperm-ethylation, *Clin. Cancer Res.*, 7, 2727, 2001.

18. Jerónimo, C. et al., Quantitative GSTP1 hypermethylation in bodily fluids of patients with prostate cancer, *Urology*, 60, 1131, 2002.

19. Ahrendt, S.A. et al., Molecular detection of tumor cells in bronchoalveolar lavage fluid from patients with early stage lung cancer, *J. Natl. Cancer. Inst.*, 91, 332, 1999.

20. Rosas, S.L. et al., Promoter hypermethylation patterns of p16, O6-methylguanine-DNA-methyltransferase, and death-associated protein kinase in tumors and saliva of head and neck cancer patients. *Cancer Res.*, 61:939–942. 2001.

21. Belinsky, S.A. et al., Aberrant methylation of p16(INK4a) is an early event in lung cancer and a potential biomarker for early diagnosis, *Proc. Natl. Acad. Sci. USA*, 95, 11891, 1998.

22. Palmisano, W.A. et al., Predicting lung cancer by detecting aberrant promoter meth-ylation in sputum, *Cancer Res.*, 60, 5954, 2000.

23. Evron, E. et al., Detection of breast cancer cells in ductal lavage fluid by methylation-specific PCR, *Lancet*, 357, 1335, 2001.

24. Jerónimo, C. et al., Detection of gene promoter hypermethylation in fine needle washings from breast lesions, *Clin. Cancer Res.*, 9, 3413, 2003.

25. Forastiere, A. et al., Head and neck cancer, *N. Engl. J. Med.*, 345, 1890, 2001.

26. Ogi, K. et al., Aberrant methylation of multiple genes and clinicopathological features in oral squamous cell carcinoma, *Clin. Cancer Res.*, 8, 3164, 2002.

27. Wong, T.S. et al., The study of p16 and p15 gene methylation in head and neck squamous cell carcinoma and their quantitative evaluation in plasma by real-time PCR, *Eur. J. Cancer*, 39, 1881, 2003.

28. Viswanathan, M., Tsuchida, N., and Shanmugam, G., Promoter hypermethylation profile of tumor-associated genes p16, p15, hMLH1, MGMT and E-cadherin in oral squamous cell carcinoma, *Int. J. Cancer*, 105, 41, 2003.

29. Chang, H.W. et al., Detection of hypermethylated RIZ1 gene in primary tumor, mouth, and throat rinsing fluid, nasopharyngeal swab, and peripheral blood of nasopharyngeal carcinoma patient, *Clin. Cancer Res.*, 9, 1033, 2003.

30. Chang, H.W. et al., Evaluation of hypermethylated tumor suppressor genes as tumor markers in mouth and throat rinsing fluid, nasopharyngeal swab and peripheral blood of nasopharygeal carcinoma patient, *Int. J. Cancer*, 105, 851, 2003.

31. Wong, T.S. et al., Promoter hypermethylation of high-in-normal 1 gene in primary nasopharyngeal carcinoma, *Clin. Cancer Res.*, 9, 3042, 2003.

32. Kresty, L.A. et al., Alterations of p16(INK4a) and p14(ARF) in patients with severe oral epithelial dysplasia, *Cancer Res.*, 62, 5295, 2002.

33. Jemal, A. et al., Cancer statistics, 2003. *CA Cancer J. Clin.*, 53, 5, 2003.

34. Naruke, T. et al., Prognosis and survival in resected lung carcinoma based on the new international staging system, *J. Thorac. Cardiovasc. Surg.*, 96, 440, 1988.

35. Soria, J.C. et al., Aberrant promoter methylation of multiple genes in bronchial brush samples from former cigarette smokers, *Cancer Res.*, 62, 351, 2002.

36. Zochbauer-Muller, S. et al., Aberrant methylation of multiple genes in the upper aerodigestive tract epithelium of heavy smokers, *Int. J. Cancer*, 107, 612, 2003.

37. Chan, E.C. et al., Aberrant promoter methylation in Chinese patients with non-small cell lung cancer: patterns in primary tumors and potential diagnostic application in bronchoalevolar lavage, *Clin. Cancer Res.*, 8, 3741, 2002.

38. Brabender, J. et al., Adenomatous polyposis coli gene promoter hypermethylation in non-small cell lung cancer is associated with survival, *Oncogene*, 20, 3528, 2001.

39. Reddy, A.N. et al., Death-associated protein kinase promoter hypermethylation in normal human lymphocytes, *Cancer Res.*, 63, 7694, 2003.

40. Brabender, J. et al., Quantitative O(6)-methylguanine DNA methyltransferase methylation analysis in curatively resected non-small cell lung cancer: associations with clinical outcome, *Clin. Cancer Res.*, 9, 223, 2003.

41. Tang, X. et al., Hypermethylation of the death-associated protein (DAP) kinase promoter and aggressiveness in stage I non-small-cell lung cancer, *J. Natl. Cancer Inst.*, 92, 1511, 2000.

42. Tomizawa, Y. et al., Clinicopathological significance of epigenetic inactivation of RASSF1A at 3p21.3 in stage I lung adenocarcinoma, *Clin. Cancer Res.*, 8, 2362, 2002.

43. Harden, S.V. et al., Gene promoter hypermethylation in tumors and lymph nodes of stage I lung cancer patients, *Clin. Cancer Res.*, 9, 1370, 2003.

44. Esteller, M. et al., Hypermethylation-associated inactivation of p14(ARF) is independent of p16(INK4a) methylation and p53 mutational status, *Cancer Res.*, 60, 129, 2000.

45. Suzuki, H. et al., Distinct methylation pattern and microsatellite instability in sporadic gastric cancer, *Int. J. Cancer*, 83, 309, 1999.

46. Issa, J.P. et al., Methylation of the oestrogen receptor CpG island links ageing and neoplasia in human colon, *Nat. Genet.*, 7, 536, 1994.

47. Toyota, M. et al., Identification of differentially methylated sequences in colorectal cancer by methylated CpG island amplification, *Cancer Res.*, 59, 2307, 1999.

48. Issa, J.P. et al., Switch from monoallelic to biallelic human IGF2 promoter methylation during aging and carcinogenesis, *Proc. Natl. Acad. Sci. USA*, 93, 11757, 1996.

49. Ahuja, N. et al., Aging and DNA methylation in colorectal mucosa and cancer, *Cancer Res.*, 58, 5489, 1998.

50. Toyota, M. et al., CpG island methylator phenotype in colorectal cancer, *Proc. Natl. Acad. Sci. USA*, 1999 Jul 20;96(15):8681–6.

51. Hawkins, N. et al., CpG island methylation in sporadic colorectal cancers and its relationship to microsatellite instability, *Gastroenterology*, 122, 1376, 2002.

52. Van Rijnsoever, M. et al., CpG island methylator phenotype is an independent predictor of survival benefit from 5-fluorouracil in stage III colorectal cancer, *Clin. Cancer Res.*, 9, 2898, 2003.

53. Nakayama, H. et al., Molecular detection of p16 promoter methylation in the serum of colorectal cancer patients, *Cancer Lett.*, 188, 115, 2002.

54. Yamaguchi, S. et al., High frequency of DAP-kinase gene promoter methylation in colorectal cancer specimens and its identification in serum, *Cancer Lett.*, 194, 99, 2003.

55. Esteller, M. et al., K-ras and p16 aberrations confer poor prognosis in human colorectal cancer, *J. Clin. Oncol.*, 19, 299, 2001.

56. Sanchez-Cespedes, M. et al., Molecular detection of neoplastic cells in lymph nodes of metastatic colorectal cancer patients predicts recurrence, *Clin. Cancer Res.*, 5, 2450, 1999.

57. Nakayama, H. et al., Molecular detection of p16 promoter methylation in the serum of recurrent colorectal cancer patients, *Int. J. Cancer*, 105, 491, 2003.

58. Lee, T.L. et al., Detection of gene promoter hypermethylation in the tumor and serum of patients with gastric carcinoma, *Clin. Cancer Res.*, 8, 1761, 2002.

59. Park, T.J. et al., Methylation of O(6)-methylguanine-DNA methyltransferase gene is associated significantly with K-ras mutation, lymph node invasion, tumor staging, and disease free survival in patients with gastric carcinoma, *Cancer*, 92, 2760, 2001.

60. Eads, C.A. et al., Epigenetic patterns in the progression of esophageal adenocarcinoma, *Cancer Res.*, 61, 3410, 2001.

61. Brock, M.V. et al., Prognostic importance of promoter hypermethylation of multiple genes in esophageal adenocarcinoma, *Clin. Cancer Res.*, 9, 2912, 2003.

62. Hibi, K. et al., Molecular detection of p16 promoter methylation in the serum of patients with esophageal squamous cell carcinoma, *Clin. Cancer Res.*, 7, 3135, 2001.

63. Wong, I.H. et al., Detection of aberrant p16 methylation in the plasma and serum of liver cancer patients, *Cancer Res.*, 59, 71, 1999.

64. Wong, I.H. et al., Frequent p15 promoter methylation in tumor and peripheral blood from hepatocellular carcinoma patients, *Clin. Cancer Res.*, 6, 3516, 2000.

65. Wong, I.H. et al., Quantitative analysis of tumor-derived methylated p16INK4a sequences in plasma, serum, and blood cells of hepatocellular carcinoma patients, *Clin. Cancer Res.*, 9, 1047, 2003.

66. Yang, B. et al., Aberrant promoter methylation profiles of tumor suppressor genes in hepatocellular carcinoma, *Am. J. Pathol.*, 163, 1101, 2003.

67. Rosty, C. and Goggins, M., Early detection of pancreatic carcinoma, *Hematol. Oncol. Clin. North Am.*, 16, 37, 2002.

68. Ueki, T. et al., Hypermethylation of multiple genes in pancreatic adenocarcinoma, *Cancer Res.*, 60, 1835, 2000.

69. Ueki, T. et al., Identification and characterization of differentially methylated CpG islands in pancreatic carcinoma, *Cancer Res.*, 61, 8540, 2001.

70. Fukushima, N. et al., Diagnosing pancreatic cancer using methylation specific PCR analysis of pancreatic juice, *Cancer Biol. Ther.*, 2, 78, 2003.

71. Yang, X., Yan, L., and Davidson, N.E., DNA methylation in breast cancer, *Endocr. Relat. Cancer*, 8, 115, 2001.

72. Lehmann, U. et al., Quantitative assessment of promoter hypermethylation during breast cancer development, *Am. J. Pathol.*, 160, 605, 2002.

73. Virmani, A.K. et al., Aberrant methylation of the adenomatous polyposis coli (APC) gene promoter 1A in breast and lung carcinomas, *Clin. Cancer Res.*, 7, 1998, 2001.

74. Evron, E. et al., Loss of cyclin D2 expression in the majority of breast cancers is associated with promoter hypermethylation, *Cancer Res.*, 61, 2782, 2001.

75. Burbee, D.G. et al., Epigenetic inactivation of RASSF1A in lung and breast cancers and malignant phenotype suppression, *J. Natl. Cancer Inst.*, 93, 691, 2001.

76. Umbricht, C.B. et al., Hypermethylation of 14-3-3 sigma (stratifin) is an early event in breast cancer, *Oncogene*, 20, 3348, 2001.

77. Virmani, A. et al., Aberrant methylation of the cyclin D2 promoter in primary small cell, nonsmall cell lung and breast cancers, *Int. J. Cancer*, 107, 341, 2003.

78. Silva, J.M. et al., Aberrant DNA methylation of the p16INK4a gene in plasma DNA of breast cancer patients, *Br. J. Cancer*, 80, 1262, 1999.

79. Silva, J.M. et al., Presence of tumor DNA in plasma of breast cancer patients: clinicopathological correlations, *Cancer Res.*, 59, 3251, 1999.
80. Hu, X.C., Wong, I.H., and Chow, L.W., Tumor-derived aberrant methylation in plasma of invasive ductal breast cancer patients: clinical implications, *Oncol. Rep.*, 10, 1811, 2003
81. Muller, H.M. et al., DNA methylation in serum of breast cancer patients: an independent prognostic marker, *Cancer Res.*, 63, 7641, 2003.
82. Pu, R.T. et al., Methylation profiling of benign and malignant breast lesions and its application to cytopathology, *Mod. Pathol.*, 16, 1095, 2003.
83. Han, M. et al., Long-term biochemical disease-free and cancer-specific survival following anatomic radical retropubic prostatectomy: the 15-year Johns Hopkins experience, *Urol. Clin. North Am.*, 28, 555, 2001.
84. Neal, D.E. and Donovan, J.L., Prostate cancer: to screen or not to screen? *Lancet Oncol.*, 1, 17, 2000.
85. DeMarzo, A.M. et al., Pathological and molecular aspects of prostate cancer, *Lancet*, 361, 955, 2003.
86. Lee, W.H. et al., Cytidine methylation of regulatory sequences near the pi-class glutathione S-transferase gene accompanies human prostatic carcinogenesis, *Proc. Natl. Acad. Sci. USA*, 91, 11733, 1994.
87. Brooks, J.D. et al., CG island methylation changes near the GSTP1 gene in prostatic intraepithelial neoplasia, *Cancer Epidemiol. Biomarkers Prev.*, 7, 531, 1998.
88. Lin, X. et al., GSTP1 CpG island hypermethylation is responsible for the absence of GSTP1 expression in human prostate cancer cells, *Am. J. Pathol.*, 159, 1815, 2001.
89. Coles, B. and Ketterer, B., The role of glutathione and glutathione transferases in chemical carcinogenesis, *Crit. Rev. Biochem. Mol. Biol.*, 25, 47, 1990.
90. Millar, D.S. et al., Detailed methylation analysis of the glutathione S-transferase pi (GSTP1) gene in prostate cancer, *Oncogene*, 18, 1313, 1999.
91. Lin, X. et al., Reversal of GSTP1 CpG island hypermethylation and reactivation of pi-class glutathione S-transferase (GSTP1) expression in human prostate cancer cells by treatment with procainamide, *Cancer Res.*, 61, 8611, 2001.
92. Nelson, W.G. et al., Preneoplastic prostate lesions: an opportunity for prostate cancer prevention, *Ann. N. Y. Acad. Sci.*, 952, 135, 2001.
93. Goessl, C. et al., DNA-based detection of prostate cancer in urine after prostatic massage, *Urology*, 58, 335, 2001.
94. Gonzalgo, M.L. et al., Prostate cancer detection by GSTP1 methylation analysis of postbiopsy urine specimens, *Clin. Cancer Res.*, 9, 2673, 2003.
95. Catalona, W.J., Beiser, J.A., and Smith, D.S., Serum free prostate specific antigen and prostate specific antigen density measurements for predicting cancer in men with prior negative prostatic biopsies, *J. Urol.*, 158, 2162, 1997.
96. Chon, C.H. et al., Use of extended systematic sampling in patients with a prior negative prostate needle biopsy, *J. Urol.*, 167, 2457, 2002.
97. Suh, C.I. et al., Comparison of telomerase activity and GSTP1 promoter methylation in ejaculate as potential screening tests for prostate cancer, *Mol. Cell. Probes*, 14, 211, 2000.
98. Harden, S.V. et al., Quantitative GSTP1 methylation clearly distinguishes benign prostatic tissue and limited prostate adenocarcinoma, *J. Urol.*, 169, 1138, 2003.
99. Harden, S.V. et al., Quantitative GSTP1 methylation and the detection of prostate adenocarcinoma in sextant biopsies, *J. Natl. Cancer Inst.*, 95, 1634, 2003.
100. Etzioni, R. et al., Overdiagnosis due to prostate-specific antigen screening: lessons from U.S. prostate cancer incidence trends, *J. Natl. Cancer Inst.*, 94, 981, 2002.

101. Maruyama, R. et al., Aberrant promoter methylation profile of prostate cancers and its relationship to clinicopathological features, *Clin. Cancer Res.*, 8, 514, 2002.

102. Tchou, J.C. et al., GSTP1 CpG island DNA hypermethylation in hepatocellular carcinomas, *Int. J. Oncol.*, 16, 663, 2000.

103. Esteller, M. et al., Inactivation of glutathione S-transferase P1 gene by promoter hypermethylation in human neoplasia, *Cancer Res.*, 58, 4515, 1998.

104. Maruyama, R. et al., Aberrant promoter methylation profile of bladder cancer and its relationship to clinicopathological features, *Cancer Res.*, 61, 8659, 2001.

105. Chan, M.W. et al., Hypermethylation of multiple genes in tumor tissues and voided urine in urinary bladder cancer patients, *Clin. Cancer Res.*, 8, 464, 2002.

106. Tada, Y. et al., The association of death-associated protein kinase hypermethylation with early recurrence in superficial bladder cancers, *Cancer Res.*, 62, 4048, 2002.

107. Chan, M.W. et al., Frequent hypermethylation of promoter region of RASSF1A in tumor tissues and voided urine of urinary bladder cancer patients, *Int. J. Cancer*, 104, 611, 2003.

108. Dominguez, G. et al., p14ARF promoter hypermethylation in plasma DNA as an indicator of disease recurrence in bladder cancer patients, *Clin. Cancer Res.*, 8, 980, 2002.

109. Valenzuela, M.T. et al., Assessing the use of p16(INK4a) promoter gene methylation in serum for detection of bladder cancer, *Eur. Urol.*, 42, 622, 2002.

110. Morris, M.R. et al., Multigene methylation analysis of Wilms' tumour and adult renal cell carcinoma, *Oncogene*, 22, 6794, 2003.

111. Battagli, C. et al., Promoter Hypermethylation of Tumor Suppressor Genes in Urine from Kidney Cancer Patients, *Cancer Res.*, 63, 8695, 2003.

112. Esteller, M. et al., Inactivation of the DNA-repair gene MGMT and the clinical response of gliomas to alkylating agents, *N. Engl. J. Med.*, 343, 1350, 2000.

113. Esteller, M. et al., Hypermethylation of the DNA repair gene O(6)-methylguanine DNA methyltransferase and survival of patients with diffuse large B-cell lymphoma, *J. Natl. Cancer Inst.*, 94, 26, 2002.

114. Taniguchi, T. et al., Disruption of the Fanconi anemia-BRCA pathway in cisplatin-sensitive ovarian tumors, *Nat. Med.*, 9, 568, 2003.

3 DNA Methylation in X Inactivation, Imprinting, and Associated Diseases

Maria Rosaria Matarazzo, Maria Strazzullo, and Maurizio D'Esposito

CONTENTS

ACKNOWLEDGMENTS

The authors gratefully acknowledge Mrs. A. Aliperti for the excellent secretarial assistance and Telethon Italy (grant no. GGP02308 to M.D.E.) for the financial support.

INTRODUCTION: AN OLD MOTIF STILL DRIVES NEW RHYTHMS

The nonrandom distribution of 5-methylcytosine in the DNA from several organisms was observed by various authors more than 40 years ago. Among them, Italian scientists from the International Laboratory of Genetics and Biophysics (LIGB) in Naples clearly demonstrated that DNA methylation was present at all the stages of early embryo development in the sea urchin, and that the 5-methylcytosine was the only minor base detectable in sea urchin embryos DNA [1,2]. Thirty years ago it was proposed that DNA methylation might have been responsible for the stable maintenance of a particular gene expression pattern through mitotic cell division [3]. Starting from these observations, many questions have been answered, yet many others still await an explanation.

What is the biological function of DNA methylation? It was proposed that DNA methylation has evolved as a generalized repression system in more complex genomes, following their size and complexity [4]. Particularly, DNA methylation seems to have evolved as a repressing mechanism against the inappropriate expression of endogenous transposons that may perturb genome organization and integrity by disrupting functional genes and/or causing chromosomal rearrangements [5]. In mammals, DNA methylation is one of the main players in determining two biological phenomena violating the dogma of genetic equality: X chromosome inactivation and genomic imprinting. Both phenomena share as a common hallmark the monoallelic expression of genes and a surprising number of epigenetic features, such as cis-acting control center, long-distance regulation, and differential DNA methylation. According to new fascinating theories, X inactivation and genomic imprinting may not share merely mechanistic aspects, but might have common origins, thus solving different biological challenges [6].

DNA methylation is accomplished through the activity of specific enzymes that transfer a methyl group to the cytosine of CpG dinucleotides. A number of DNA methyltransferases (DNMTs) with different roles have been identified in mammals. DNMT3A and DNMT3B are mainly devoted to the *de novo* methylation, whereas DNMT1, with preferential activity for hemimethylated DNA, acts mainly as maintenance methyltransferase. Null mutations of the three DNMTs are lethal [7,8], clearly demonstrating that DNA methylation is essential for mammalian survival.

How could the repressive signal be interpreted? DNA methylation can directly interfere with the consensus sequence of trans-acting factors. In addition to repulse factors, DNA methylation can attract different ones, such as the so-called methyl binding proteins (MBDs), a family of five proteins that share a common motif, the methyl binding domain [9]. These proteins seem to convert the DNA methylation signal into a repressive state of chromatin, recruiting large complexes that include histone deacetylases [10]. A suggestive hypothesis proposed that to increase the fidelity of DNA methylation silencing during vertebrate evolution, an increase arose in the number and functional diversity of the methyl binding proteins [11].

In the following paragraphs, we discuss the molecular details of the involvement of DNA methylation in the regulation of X inactivation and of genomic imprinting. On the other hand, as was recently and excellently reviewed [12,13], disturbances of the regulation by DNA methylation is expected to give rise to genetic syndromes. What can we learn from them in terms of molecular mechanisms? In this chapter we will not try to give an answer, but offer a review of the current knowledge.

DNA METHYLATION AND
X CHROMOSOME INACTIVATION

The involvement of DNA methylation in X chromosome inactivation dates back to 1975, when Riggs predicted the "maintenance methylase" model for somatic inheritance of methylation patterns, suggesting that DNA methylation would play a role in mammalian X inactivation and development. He emphasized that especially for X inactivation, which was thought to be based on specific protein–DNA interactions, it was unnecessary to involve DNA sequence changes because protein binding was sensitive to methylation. Also, methylation patterns would be clonally heritable given a maintenance methylase. A maintenance methylase was defined as a DNA methyltransferase that preferentially acts on hemimethylated sites [3]. Since then, DNA methylation has long been considered an important component of X inactivation, particularly as a mechanism for steadily maintaining the inactive state and for transmitting it as a "memory" through cell generations.

General Features of X Inactivation

X chromosome inactivation is the process whereby female mammals transcriptionally silence one of their two X chromosomes to compensate for having a double dosage of X-linked genes as compared to males. This is necessary because the unbalanced gene dosage of sex chromosomes between males and females represents an impediment to normal development. Several mechanisms have been adopted by different species to circumvent this problem [14]. The distinctive feature of mammalian X inactivation, compared to other dosage compensation strategies, is that silencing of one X chromosome takes place while its homologue in the same nucleus remains genetically active. Visible as a compact structure at the nuclear periphery (the Barr body), the inactive X is similar to a constitutive heterochromatin for the following features: hypermethylation within CpG islands of house-keeping genes, late replication during S-phase, and association with modified nucleosomes. These nucleosomes are peculiarly composed of hypoacetylated and methylated histone H3, hypoacetylated histone H4, and enriched with H2A variants [15]. But the feature making the inactive X unique among the chromosomes is that it is associated with a non-coding RNA molecule, the *Xist* RNA (X-inactive specific transcript) [16] over its entire length. Although it is not completely clear how each of these characteristics contributes to the X inactivation process, they seem to act synergistically [17].

The onset of X inactivation is triggered upon cellular differentiation during early embryogenesis in the mouse, occurring shortly after the implantation of embryos. In the first differentiated tissue, trophectoderm and primitive endoderm, X inactivation occurs early, whereas in the tissues that form the embryo proper, it occurs a little later. An even later site of X inactivation is in the male germline during spermatogenesis. Invariably, it is always the paternal X that is inactivated in extraembryonic lineages (imprinted inactivation) [18]; in contrast, either the maternal or the paternal X can be inactivated in the cells forming the embryo proper (random inactivation). Thus, in mammals, except marsupials where only imprinted paternal inactivation is observed, two different forms of X inactivation are described; regardless of the mechanism, the final result, in terms of dosage compensation, is identical. A possible third mechanism may operate during spermatogenesis, as spermatocytes are the only cells believed to inactivate the single X in males.

Xist and *Tsix*: The Antagonist RNAs

A major contribution to our current understanding of X inactivation has come from classical genetic studies: whether one X remains active or becomes silenced depends on the activity of an X-linked central control that, through a counting and a choice step, coordinates silencing between the two X chromosomes and initiates it. This small region, called the X inactivation center (Xic) [19], harbors several unusual genetic elements. Among them, the pivotal element is *Xist*, a gene expressed only from the inactive X and whose nuclear RNA product forms a "cloud" on it [20]. Accumulation of *Xist* RNA could then recruit the silencing complexes that establish a heritable chromatin conformation along the length of the X chromosome. Results from gene knockout and transgenic studies have widely confirmed this model [21] and the key role of *Xist* RNA. Following a recent, finer mutational analysis, the specific domain responsible for silencing has been identified within a repeat motif at the 5 end of *Xist* RNA, thus indicating a modular nature for this RNA [22].

A molecular breakthrough in understanding the counting and choice steps of X inactivation came with the discovery of *Tsix*, an overlapping antisense RNA of *Xist*, which, like *Xist*, is a noncoding RNA [23]. Its expression pattern is dynamically associated with *Xist* expression during the process of X inactivation. Prior to the onset of X inactivation, *Xist* is expressed at low level from each X chromosome, together with *Tsix*. When X inactivation initiates, *Xist* transcript levels are upregulated from the chromosome that will become the inactive X (Xi), while *Tsix* expression, which has a repressive role, is lost but persists on the future active X (Xa). How the *Tsix* expression blocks *Xist* RNA accumulation at the molecular level remains unknown. Two main potential mechanisms, which are not mutually exclusive, have been proposed: an RNA interference mechanism that destabilizes *Xist* RNA and the silencing the *Xist* promoter. They, in fact, may work together to repress *Xist* expression [24]. It has been demonstrated that *Tsix* also regulates X chromosome choice. A cis-acting center for choice and imprinting (ICR, imprinting control region) of inactive X lies at the CpG island of *Tsix* promoter (at the 3 end of *Xist*), including a repeat element known as DXPas34, as its deletion abolishes random choice in cells of the embryo proper to favor inactivation of the mutated X and disrupt maternal *Xist* imprinting in extraembryonic tissues [25,26]. Thus, while imprinted X inactivation is paternally directed and random X inactivation is zygotically controlled, both work through *Tsix* to regulate *Xist*. Random X inactivation in embryonic tissues is probably accomplished by the erasure of an imprinting mark that controls the imprinted expression of *Xist* and *Tsix*. Differential DNA methylation of the *Xist* promoter could be an obvious candidate for such an imprint, but detailed bisulfite sequencing has failed to confirm early reports based on PCR-restriction enzyme analysis suggesting methylation differences in female germline and early embryos [27]. Given the importance of the CpG island of *Tsix* in counting and choice, another attractive target for this imprint might lie there. Although DXPas34 is differentially methylated on the Xa and Xi of somatic female cells [28], again bisulfite sequencing in oocytes and spermatocytes failed to find any differences in the methylation pattern between them [29]. Thus, whether DNA methylation is the imprinting mark erased in the embryonic lineage to allow random X inactivation remains undetermined.

A striking role for CpG and non-CpG methylation of the DXPas34 region has been recently proposed in the model of chromosome choice invoking trans-acting blocking factors that protect the future Xa from silencing, first postulated by Lyon in 1996 [30]. Intriguingly, the switch for *Xist* action might depend not only on the activity of its antisense counterpart, *Tsix*, but also on the binding of CTCF (CCCTC-binding factor), a zinc-finger transcription factor that has been implicated in boundary formation and chromatin insulation [31]. Latest models hypothesize that CTCF binds the regulatory element, DXPas34, within *Tsix* whose transcription then marks and protects the future Xa from inactivation. Conversely, on the future Xi, *Tsix* must be extinguished so that *Xist* RNA accumulates along the chromosome. Allelic methylation differences within the proposed differentially methylated region in this regulatory element would enable discrimination of the X chromosomes and binding of CTCF to only one allele of the *Tsix*. Suppression of *Xist* by binding of the CTCF to the Xa could be achieved by direct activation of its repressor, *Tsix*, or by blocking access to a putative enhancer located downstream (Figure 3.1A). Although

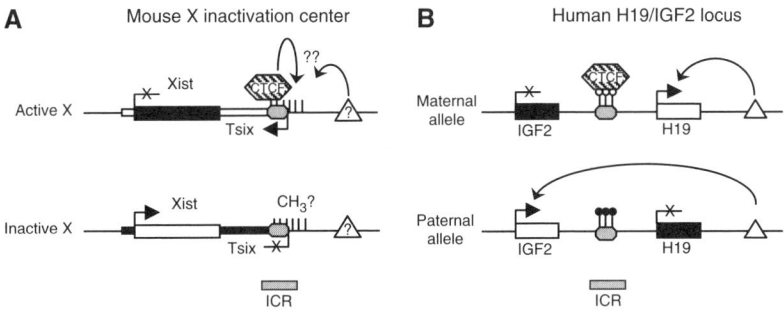

FIGURE 3.1 CTCF binding sites at the mouse Xic (**A**) and at human 11p15.5 DMR1 (**B**). □ expressed genes, ■ repressed genes, ● CTCF binding sites, ⑪ unmethylated DMR, ⑪ methylated DMR, ⅠⅠⅠ CpG island, Δ enhancer. (Modified from: Lee, J.T., *Curr. Biol.*, 13(6), R242–254, 2003.)

methylation blocks CTCF *in vitro*, whether DNA methylation might regulate its binding *in vivo* remains unclear because, as mentioned, conflicting evidence has been reported for the methylation status of the DXPas34 region. Because CpG methylation has relatively little effect on CTCF binding to DXPas34 *in vitro*, but non-CpG methylation abolishes CTCF binding in this assay, the searching for an imprint in the form of differential non-CpG methylation at DXPas34 would be the next step. Without DNA methylation, the CpG island may regulate the *Tsix* gene through other chromatin-associated changes. Whatever will be the case, the model proposed would provide a fascinating solution to the long-standing dilemma of how a cell could discriminate between two seemingly equivalent X chromosomes, leading them to divergent fates [32].

SILENCING FACTORS: PROTEIN NUCLEIC ACID WANTED

One important feature of X inactivation is its spreading from the Xic, where this process initiates, to the entire X chromosome. Discovering the molecules that are involved in this process is the goal of an intensive research effort. At present, almost nothing is known about the partners, proteins, or nucleic acids mediating the binding of the *Xist* transcript to the X chromosome or about the chromatin remodeling that is induced by this association. One approach to identify the proteins recruited by *Xist* RNA has been to analyze the temporal order of events leading to stable X inactivation after expression of *Xist* in differentiating XX embryonic stem (ES) cells (Figure 3.2A). In this *in vitro* model system, global H4 hypoacetylation, recruitment of the variant histone, macroH2A, and CpG methylation are all relatively late events in the X inactivation process, occurring well after *Xist* RNA coating and transcriptional silencing of genes. Rather than being involved in the establishment of the inactive state, these epigenetic marks appear instead to be involved in maintaining it.

Because there is a tight developmental window during which cells are both responsive to and dependent upon *Xist* expression [33], it is suggested that the recruitment of factors that mediate this transcriptional shutdown is developmentally regulated within a defined window of time. It is possible that either the partners of *Xist* mediating silencing are no longer present or competent or that the chromatin of the X chromosome has become remodeled to resist their effects. In this latter scenario, irreversibility might reflect the arrival of promoter DNA methylation, as recently postulated [10]. The implication that irreversibility implicates DNA methylation is supported by the frequent reactivation of a X-linked transgene in mouse embryo cells and in cultured somatic cells when Dnmt1 is absent or inhibited [34].

The relative instability of X-inactivated genes in the marsupials and the tendency for cells of the human chorionic villus to reactivate X-inactivated genes have also both been linked to hypomethylation of CpG islands of X-linked genes [35,36]. In addition, X inactivation can be relieved by treatment of somatic cells with demethylating agents [37], demonstrating that methylation clearly contributes to reinforce inactivation.

The stepwise process leading to establishment of X inactivation in differentiating XX ES cells suggests that it might be possible, by determining the earliest detectable events, to identify candidates for the silencing factors recruited by *Xist* RNA. Recent

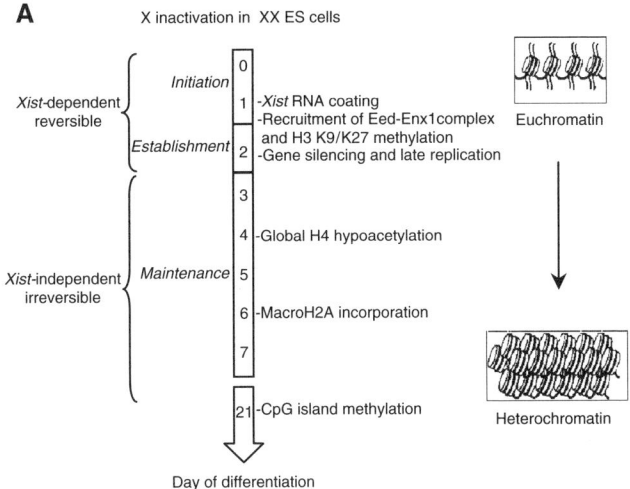

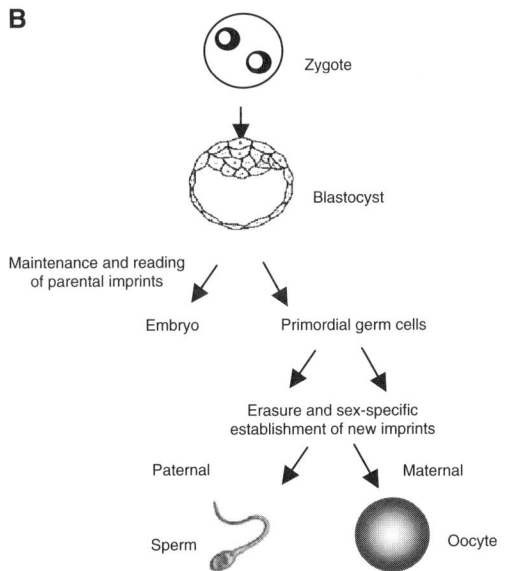

FIGURE 3.2 (**A**) Timing of epigenetic events in initiation, establishment, and maintenance of X inactivation during female ES cell differentiation. (**B**) Lifecycle of imprints during mammal development.

studies have focused attention on modifications of the histone H3 N-terminus. If DNA methylation is a late event, histone methylation at H3-lysine 9, after a previous deacetylation, as well as loss of methylation at H3-lysine 4, seem to be very early steps of X inactivation. Furthermore, the late appearance of H3 methylation at the promoters of X-linked genes suggests that it may also be involved in "locking in" the transcriptional inactivity of X-linked genes [38]. The specific interplay between the two kinds of epigenetic modifications, DNA and histone methylation, in gene silencing is still under investigation. It will be interesting to find out their relative contribution to the silencing complex involved in X inactivation. Another event temporally regulated is the recruitment to the inactive X of Polycomb group complex, Eed-Enx1. Based on the latest models, the function of this complex would be to establish, interacting with *Xist* RNA, histone H3 methylation at lysines 27/9 very early after a previous deacetylation by a histone deacetylase (HDAC) yet unidentified, but also bound to *Xist* [39].

The involvement of other factors, such as those that bring about DNA methylation in the initiation and propagation of X inactivation, however, remains to be investigated. Early results had suggested a role for DNA methylation not only in the maintenance of inactive state of X-linked genes but also in an initiation step, namely in the controlling *Xist* expression during the enactment of the active X. Differential DNA methylation has been described for the *Xist* promoter on the active X and inactive X in somatic cells and in undifferentiated ES cells [40]. Moreover, differentiating ES cell lines lacking the Dnmt1 exhibit, in fact, *Xist* de-repression in some cells and initiate inactivation of X-linked genes in cis [41], indicating a requirement for DNA methylation in blocking initiation of silencing in ES cells. More recently, studies of Dnmt1-knockout mice indicate that the zygotic function of Dnmt1 might not be essential for the establishment of X inactivation in embryonic and extraembryonic tissues [34]. The presence of maternal methyltransferases might be sufficient to allow the initiation and spreading of X inactivation to proceed. Analyses of Dnmt3a and Dnmt3b protein levels in mice suggest that *de novo* methylation at the time of X inactivation is largely due to the Dnmt3b [42], thus implicating this enzyme in the CpG island methylation. Hypermethylation might recruit a methyl-CpG-binding protein, such as MECP2 and MBD2, which in turn would recruit chromatin-remodeling complexes.

Useful insights on the role of DNA methylation in spreading step X inactivation came from studies carried out in human ICF (immunodeficiency, centromeric decondensation, and facial anomalies) patients who have DNMT3B mutations [43]. In these subjects several genes have their CpG island hypomethylated, and among them only two are inactivated imperfectly [44]. Because X inactivation in ICF cells is basically normal at the cytological level, and mostly at the individual gene level, the establishment and spread of X inactivation appear to be normal in a DNMT3B-deficient background. This is not surprising because, as mentioned, evidence shows methylation of CpG islands as a secondary silencing event.

Recently, a report analyzed the LINE-1 (L1) repeats in ICF cells and their methylation status [45]. This is particularly interesting because of implications for the so-called Lyon repeat hypothesis. Mary Lyon proposed that L1s might function to promote spreading along the X chromosome, acting as the "way station" or "booster elements" originally hypothesized by Riggs, probably through an *Xist*-mediated intrachromosomal pairing of L1 repeats [46]. These elements were interesting candidates for this function in view of the L1-rich nature of the human and mouse X's compared with the autosomes. By analyzing L1 CpG islands of the inactive X chromosome in ICF patients, Hansen found that they are hypomethylated.

Because methylation inhibits recombination and other processes that require pairing, at the time of X inactivation L1 repeats are probably unmethylated [45]. Thus, a role in the spreading of X inactivation could be played by unmethylated L1 elements rather than by methylated L1s.

A MOLECULAR LINK BETWEEN X INACTIVATION AND GENOMIC IMPRINTING?

Imprinted X inactivation is also mediated by *Xist* and *Tsix* RNA as random X inactivation. However, a distinct mechanism regulates *Xist* expression, so that the transcription occurs only on the paternally derived X chromosome, as well as the downstream events. The role of DNA methylation, for example, is less clear than in random X inactivation. Indirect observations indicate that DNA methylation of CpG islands might not occur [34,47]. According to this, lower levels of methylation at constitutive heterochromatin have been seen in these lineages relative to the embryo proper [48].

Although imprinted X inactivation is believed to be the ancestral mechanism of dosage compensation in mammals, already observed in all tissues of marsupials, the basis of its existence is still mostly enigmatic. An old point of view has been very recently reexamined by Huynh and Lee. These authors have demonstrated that the paternal X is actually transmitted to the zygote as a preinactivated chromosome. In this model, the paternal X chromosome exhibits a gradient of silencing already at the zygotic gene activation (two-cell stage) probably deriving from the paternal germline [49]. Recent evidence is in contrast to these conclusions, leaving the issue still debatable. Namely, it has been reported that the male germline does not confer any imprints to the paternal X chromosome, as it appears to be active at the two-cell stage and becomes inactive only after *Xist* RNA coating occurring from the four-cell stage onward (see Okamoto, I. et al., *Science*, 303, 644–649, 2004). Regardless of the detailed molecular aspects of the imprinted X inactivation, the existence of this phenomenon may provide the evolutionary link between X inactivation and autosomal imprinting [6].

DNA METHYLATION AND GENOMIC IMPRINTING

Genomic imprinting is a system of non-Mendelian inheritance that is unique to eutherian mammals, marsupials, and higher plants. It is an epigenetic mechanism of gene regulation that determines the parent-of-origin-dependent expression of a number of genes during development. Imprinted genes are marked in the male and female germline and retain molecular memory of their parental origin, resulting in allelic expression differences. DNA methylation is a basic mechanism for establishment and maintenance of the imprinting status.

The concept of genomic imprinting as a possible mechanism to explain functional differences between maternal and paternal genomes in mammals was proposed in 1984 [50,51]. These two groups demonstrated that gynogenetic or androgenetic embryos that had either two maternal or paternal pronuclei showed early embryonic lethality and never developed to term. These early observations were largely sustained by subsequent numerous genetic and molecular demonstrations. A few years later, a role of imprinting in human diseases (Prader-Willi syndrome, [52]) was suggested. In the early 1990s, the first imprinted gene was identified (IGF2, [53]). Systematic screening methods have contributed to the identification of a large number of imprinted genes and to the precise localization of imprinted chromosomal regions. To date, about 60 imprinted genes have been isolated from the human and mouse genomes. Using restriction landmark genome scanning (RLGS), it has been estimated that there are roughly 100 imprinted genes in the mouse genome [54]. Although some of the genes analyzed are imprinted in mice but not in humans, most of the imprinted genes are conserved. The mouse chromosomal region, responsible for the embryo development failure, was mapped to specific chromosomes by the use of artificially created uniparental disomies (UPD), in which two copies of a gene were inherited entirely from a parent.

Several hypotheses have been proposed to explain the evolution of a particular regulative mechanism of gene expression. The more generally accepted is the parental conflict hypothesis. According to this, evolutionary force that propagates genomic imprinting in mammals might be a parental conflict for the maternal resources, promoting imprinted expression of genes that function in growth regulation [55]. Few hypotheses deal with mechanistic aspects and the molecular basis of imprinting: (1) It has been suggested that repeats and retrotransposons might be marked and silenced by allele-specific epigenetic modifications; thereby a host defense mechanism may have provided the molecular basis for genomic imprinting [56]. (2) Imprinting might be evolved from incomplete gene silencing, thereby leading first to random monoallelic gene expression and later to a specific silencing [56]. (3) Duplication and translocation of chromosomal segments might play a role in imprinting establishment through a different possible mechanism such as the necessity of a dosage compensation to silence additional gene copies [55]. (4) As mentioned, imprinting and X chromosome inactivation may have a common origin [6].

COMMON FEATURES OF IMPRINTED GENES

The mammalian genes that undergo parental-specific regulation and the chromosomal region where they map share an interesting range of physical, genetic, and epigenetic features that may be summarized in some main relevant points.

Imprinted genes are rarely found dispersed; around 80% of them are physically linked in clusters with other imprinted genes. The clustered organization of imprinted genes is thought to reflect coordinated regulation of the genes in a chromosomal domain. Imprinting centers or imprinting control elements (ICs) have been discovered in some clusters; the ICs seem to be necessary for the regional control of imprinting and imprinted expression.

TABLE 3.1
A Schematic Summary of the Main Genetic Abnormalities in Prader-Willi, Angelman, and Beckwith-Wiedemann

Molecular Class	Prader-Willi Syndrome	Angelman Syndrome	Beckwith-Wiedemann Syndrome
Deletion or duplication	Paternal deletions 75%	Maternal deletions 75%	Paternal duplication <1%
Uniparental disomy	Maternal (meiotic) 22%	Paternal meiotic 2%	Paternal somatic 10%
IM: Inherited	IC microdeletions 1%	IC microdeletions 1%	None
Sporadic	" 2%	" 2%	LOI at IGF2 (20%) or LIT1 (50%)
Translocations, chromatin effects	Paternal inheritance (<1%)	—	Maternal inheritance <1%
Gene mutations	—	Maternal UBE 3A 7%	Maternal CDKN1C 10%
unknown	—	13%	10–20%

Source: Adapted from Nicholls, R.D., *J. Clin. Invest.*, 105(4), 413–418, 2002.

Deletions and inappropriate methylation of these regions are associated with loss of imprinting and human diseases (see Table 3.1).

The comparison of proteins codified by different imprinted genes shows no interesting common features. It is possible to underline only functional similarities between some proteins with a role in fetal growth and development. Instead, the analyses of nucleotide sequences show two elements of functional significance. First, they are unusually rich in CpG islands (in the mouse, 88% of imprinted genes are associated with CpG islands vs. the 47% in the whole genome). Second, the CpG islands are frequently surrounded by direct repeats. The repeats, together with the CpG islands associated with them, are involved in conferring or maintaining differential methylation of the imprinted allele [54].

The parental alleles of an imprinted gene show differences in the methylation pattern. As described below, the differentially methylated regions (DMRs) can have different properties and methylation patterns during development and in germ and somatic cells. Deletion of a DMR results in loss of imprinting. DMRs are generally CpG rich and often fulfill the criteria of CpG islands. Some DMRs (e.g., H19, IGF2R) contain also repeat elements.

Imprinted genes can differ with respect to bulk chromatin structure for specific modification, such as the accessibility of DNA to transcriptional machinery and histone covalent modifications. Methylation imprints are associated with chromatin imprints [57].

Imprinted genes often reside in chromosomal regions that undergo asynchronous replication [58]. Meiotic recombination frequency of these segments may differ in male and female germ cells [59].

Within the imprinted gene clusters, genes that encode for untranslated RNA and antisense RNA possibly involved in imprinted expression control have been identified [60].

MOLECULAR BASES OF IMPRINTED EXPRESSION REGULATION

The mechanism of genomic imprinting is complex and not completely understood. It is well established that DNA methylation plays a leading molecular role. Cytosine methylation marks imprinted genes differently in gametes, and the inheritance of these epigenetic marks leads to differential gene expression. Usually methylation marks the unexpressed allele, but it is not an absolute rule. Genomic imprinting changes during the life cycle of the organism and proceeds in a different way in germ and somatic cells (see Figure 3.2B). In somatic cells, imprinting may be established in a tissue-specific manner.

Imprints are established during the maturation of germ cells to sperm or eggs. After fertilization, the differential mark of paternal and maternal alleles in somatic cells is maintained during organism development. In the germ cells of the new organism, imprints are erased at an early stage in the primordial germ cells, and at a later stage of development, in inner cell mass, they are established again in a sex-specific manner. Imprint determination during development may be schematized in three main steps: erasure, establishment, and maintenance.

Erasure. In mouse germ cells during erasure, there is a marked and apparently genome-wide demethylation. First demethylation occurs in male pronucleus and seems to be independent on DNA replication. After the formation of the zygote, at the stage of blastocyst, both paternal and maternal chromosomes undergo progressive demethylation erasing most, but not all, of the marks that are inherited from the gametes. Whether this mechanism is passive or active is still unknown. Whether the methylation marks on imprinted genes are protected from genome-wide demethylation is also discussed [57]. The general DNA demethylation is accompanied by chromatin modifications that seem to facilitate the genomic reprogramming.

Establishment. The exact timing of the imprints' resetting is not yet clear and seem to be quite different in oocytes and sperm [61]. After erasure, *de novo* methylation of cytosine begins in both sex germlines at late fetal stages and continues after birth. The maternal or paternal imprints established in germ cells lead to the formation of the primary DMRs. *De novo* and maintenance methyltransferases seem to be needed for remethylation in sperm and oocyte. Dnmt1 and its germ-cell-specific isoforms (Dnmt1o) are candidates; but also the known somatic *de novo* methyltransferase, Dnmt3a and Dnmt3b, may carry out the same function in germ cells [60]. It is also yet unclear how

Dnmts specifically target DMRs in eggs and sperms. This phenomenon implies the existence of germline-specific factors that allow the recognition of the specific DMRs and the sex of the lineage. This idea is well supported by the evidence that deficiency of the Dnmt1 causes a postzygotic loss of imprinting that cannot be restored by Dnmt1, Dnmt3a, and Dnmt3b. Putative candidates for this function may be direct repeats [54], as repeated elements are often involved in local heterochromatization [62].

Maintenance and reading. During mammals' development a wave of genome-wide demethylation after fertilization and a wave of *de novo* methylation after implantation exist that imprinted DMRs have to resist. How DMRs enact this defense is still under discussion. The regulation of imprinted status also involves histone acetylation and possibly histone methylation [57]. When the methylation and chromatin imprints are established, they have to be conversed in differential transcription. Reading mechanisms better explored are:

Differential silencing by CpG island or promoter methylation.

Regulation by antisense transcripts associated with CpG island or promoter methylation (i.e., Air gene).

Regulation by silencing factors that repress the promoter in cis.

Allele-specific regulation of neighboring genes by differential methylation of boundary elements within a CpG island. A well-described example is the role of CTCF in recognition of an unmethylated boundary in the H19/IGF2 system [60]. The same binding sites are recognized by Boris [63], another protein that seems to act in a tissue-restricted manner (testis). Recent reports suggest that Boris and CTCF could serve as counteracting genes by manifesting and reprogramming states, respectively [64].

IMPRINTING IN GENETIC DISEASES

Genomic imprinting plays a critical role in embryogenesis as evidenced by certain aberrations of human pregnancy, such as hydatiform mole or ovarian dermoid cysts. A large number of imprinted genes identified in humans and mice are involved in pre- and postnatal development. Recent studies [65] associate imprinted genes with determining brain development, and two paternally expressed genes have been associated with the regulation of maternal behaviors in mice. Evidence suggests that imprinted mammalian genes influence complex neuropsychological behavior and neuropathological disorders [66]. In the following, three of the most investigated syndromes are summarized.

PRADER-WILLI SYNDROME AND ANGELMAN SYNDROME

Prader-Willi syndrome (PWS, OMIM #176270) and Angelman syndrome (AS, OMIM #105830) are autosomal dominant disorders showing parent-of-origin effects, since the inherited diseases are transmitted from only one of the parents. These two clinically distinct genetic diseases are associated with imprinting on chromosome 15q11-q13 and have a frequency of about 1:10,000/1:40,000. Both syndromes are neurodevelopmental disorders associated with deficiencies in sexual development and growth; PWS and AS patients show behavioral and neurological problems including several learning disabilities. Major diagnostic criteria for Prader-Willi syndrome include hypotonia, hyperphagia, marked obesity, hypogonadism, and developmental delay. Major diagnostic criteria for AS are ataxia, tremulousness, sleep disorders, hyperactivity, and a peculiar happy disposition with outbreaks of laughter.

The most common genetic feature associated with Prader-Willi and Angelman syndromes (see Table 3.1) is an interstitial *de novo* deletion of 3–4 Mb in the chromosome 15q11-q13. An unequal crossing-over between low copy repeats (duplicons) present in the region seems to be the main cause of these deletions. The duplicons contain the HERC2 gene, a highly conserved gene that, when mutated in mice, affects development and fertility [67]. These deletions are of maternal origin in AS patients and of paternal origin in PWS patients. Imprinted expression is coordinately controlled by an IC, which is functional in the germline and in early postzygotic development. The IC regulates the establishment of parental-specific allelic differences in DNA methylation, chromatin structure, and expression [68,69]. The 15q11-q13 IC has a bipartite structure. One part seems to be involved in switching of paternal imprinting in maternal; the other seems to be responsible for the opposite switch [70].

While the main genes involved in AS have been identified, this goal for PWS is still unaccomplished. Recent evidence supports the existence of at least 30 presumed genes in the region [71]. Several genes have been identified and characterized, and a paternal expression was demonstrated. At least two genes in this region, UBE3A and ATP10C, are maternally expressed in some tissues. Deletions and other alterations (see Table 3.1) in UBE3A have been shown to cause AS. The UBE3A gene encodes for a ubiquitin protein ligase, which is thought to play a role in marking, by ubiquitination, proteins destined for degradation within the brain. This gene is expressed in the hippocampus and Purkinje cells with a brain-specific imprinting (maternal allele expression) while it is biallelically expressed in other tissues [72].

In contrast to AS, which is often associated with single gene mutations, PWS is always associated with chromosomal abnormalities affecting multiple genes. The finding of many paternally imprinted genes has contributed to the identification of several candidate genes for PWS. The SNRF/SNRPN locus is a prime candidate. It is notable that the 5 end of the SNURF/SNRP maps in the IC. This locus is highly complex with a long variable policistronic precursor transcript with multiple functions. At least four transcripts are associated with this locus: SNURF, IPW, PAR5, and multiple small nucleolar RNAs (snoRNAs) [73]. The snoRNAs have been proposed to confer most of the PWS phenotype [74].

BECKWITH-WIEDEMANN SYNDROME

Beckwith-Wiedemann syndrome (BWS, OMIM #130650) is an autosomal multigenic syndrome with a clear parent-of-origin phenotype. This syndrome has a mainly sporadic inheritance and, only in a few cases (CDKN1C gene mutations), it is autosomal dominant. Clinically it is characterized by pre- and postnatal overgrowth, macroglossia, and anterior abdominal wall defects. Additional, but variable, complications include organomegaly, hypoglycemia, genitourinary abnormalities, and, in about the 5% of affected children, embryonal tumors, frequently Wilm's tumor (in BWS patients, the rate of Wilm's tumor is over 1000 higher than in the normal population). As for the other syndromes described, the molecular bases of BWS are very complex. The 11p15.5 is organized into two imprinted domains in which genomic imprinting is controlled by separate imprinting control regions. The genetic mechanisms underlying BWS (see Table 3.1) imply a role either for paternally and maternally expressed genes. Oversimplifying, in the BWS pathogenesis two main mechanisms seem to be involved: paternally derived duplications, UPD, and loss of imprinting (LOI) at insulin-like growth factor (IGF2) locus; and maternally derived translocations and specific mutations at the CDKN1C (p57^{KIP2}) locus [75]. These two genes contribute to the BWS phenotype together with a number of other imprinted or nonimprinted genes, which are localized in the two domains [76].

The distal domain 1 contains the imprinted genes IGF2, H19, and INS, and a differentially methylated region, DMR1, IGF2, and H19, are closely linked and oppositely imprinted. IGF2 encodes a paternally expressed fetal growth factor; H19 encodes an untranslated pol III transcript. The expression profile of IGF2, in normal development, resembles the spectrum of tissues and organs involved in BWS. In the mouse, these two genes share, and seem to compete for, a cluster of enhancers located at the 3 end of the H19 gene. The DMR1 is located 2 kb upstream of H19 and regulates reciprocal imprinted expression of IGF2 and H19. This regulative region functions as a chromatin boundary or insulator, which is recognized by CTCF (see Figure 3.1B). DMR1 is methylated on the paternal chromosome while it is unmethylated on maternal allele, and is so accessible to the binding of CTCF. This binding blocks IGF2 promoter interaction with downstream enhancers and favors H19 transcription.

The proximal domain 2 contains several imprinted genes. A particular role in BWS syndrome is recognized for CDKN1C, a cyclin-dependent kinase inhibitor, which is maternally expressed. Although this gene is involved in cell growth regulation, it is rarely involved in tumors. Mutations in this gene do cause BWS and are associated with the dominant transmission of the syndrome. The other genes identified are IPL/TSSC3, maternally expressed; IMPT1/TSSC5, maternally expressed; KCNQ1, maternally expressed and encodes for a number of transcripts. One of these codifies for a part of a potassium channel, which is associated with other human disorders (long QT syndrome and Jervell-Lange-Nielson syndrome) [77]. Aberrations in this gene are strongly associated with BWS. The intron 10 of KCNQ1 contains DMR2. The paternal DMR2 is not methylated and permits the transcription of a long antisense transcript known as KCNQ1OT1 or LIT1. DMR2 seems to have insulator activity in mice [78] and insulator and silencer activity in humans [75].

DNA METHYLATION DISORDERS AND HUMAN DISEASES

The list of human diseases caused or involving methylation disturbances is quite long. These pathologies are often syndromic, given the pleiotropic effect of mutations on the phenotype. It is possible to subdivide these defects according to (1) perturbation of the methylation level; (2) failure of the interpretation of the methylation signal; and (3) inappropriate methylation of regulatory regions.

In addition to germline mutations, somatic effects may also perturb DNA methylation control and, subsequently, gene expression, possibly predisposing to a disease. This is the case of the hyperhomocysteinemia secondary to chronic renal failure [79]. Additionally, even in striking contrast to the current notion of erasure of the DNA methylation mark after fertilization discussed before, epigenetic regulation, such as DNA methylation, could play a role in predisposing multifactorial defects, such as diabetes [80,81].

ICF SYNDROME: IS IT A MATTER OF (METHYLATION) LEVEL?

Disturbance of the methylation level is seen in a rare autosomal-recessive disease, ICF syndrome (ICF, OMIM #242860). ICF patients invariably present two peculiar signs: a severe immunodeficiency and a peculiar decondensation of the pericentromeric heterochromatin, especially of chromosomes 1 and 16, and less frequently chromosome 9. Other variable signs at a clinical level are facial anomalies and psychomotor and mental retardation that seem hallmarks for methylation-related diseases. The syndrome was genetically mapped on chromosome 20 by genetic analysis [82], and the causative gene was identified as that encoding DNMT3B [8,43,83]. Mutations usually reside in the carboxy terminus, containing the catalytic domain, leading to the hypothesis that such loss of function is responsible for the disease [84]. What are the target sequences of DNMT3B enzyme activity? The hypomethylation of the genome in ICF syndrome is much less generalized than after treatment with DNA methyltransferases inhibitors, such as 5-azacytidine. In addition to the hypomethylation of the major satellite [85], an abnormal pattern of the facultative heterochromatin belonging to the inactive X chromosome has been shown. The result was unexpected since there are no phenotypic differences between male and female ICF patients [86]. However, it does not prevent the occurrence of X inactivation [87]. A number of repetitive sequences are affected by DNMT3B loss of function; the hypomethylation of nonsatellite sequences, such as D4Z4 and NBL2, after a whole genome scan for methylation differences in ICF patients has been reported [88]. A detailed study of L1 methylation in ICF syndrome [45] has also been published. L1 elements are hypermethylated, both on the active and inactive X in normal cells; in ICF patients' derived cells, DNMT3B loss of function produces no effect on the Xa but a significant hypomethylation of Xi-specific LINEs. This is a unique phenomenon of identical allele modification performed by different DNA methyltransferases.

What about genes affected in ICF syndrome? This argument is still in debate. In earlier experiments, no demethylation was observed in the single DNMT3B mutant, whereas the double mutant DNMT3A/3B shows demethylation of the 5 region of *Xist* and the DMR2 of IGF2 genes. Hansen and coworkers showed a generalized hypomethylation of X-inactive genes, with a gene-specific reactivation seen only for G6PD and for the pseudoautosomal gene SYBL1. Such specific biallelic expression led to an advancement of replication timing [44]. SYBL1 gene histone code was further investigated in an ICF patient cell line: DNA hypomethylation is accompanied by histone H3/4 hyperacetylation and differential methylation of H3K4 and H3K9, with enrichment of the former and a decrease of the latter [89]. This phenomenon is probably mechanistically linked with MBD binding of SYBL1 promoter, which is also affected in ICF syndrome cell lines (Matarazzo, M.R., et al. in progress).

Since the main system affected in ICF syndrome is the immune system, the study of Ehrlich's and Hanash's teams to derive the expression profiling of these patients is of extreme interest. Over 30 genes were found deregulated, many of them lymphogenesis associated, with specialized function in lymphocyte signaling, maturation, and migration [90].

RETT SYNDROME: LOST IN "TRANSCRIPTION"

Rett syndrome (RTT, OMIM #312750) is a neurodevelopmental disorder of postnatal brain growth constituting one of the most common causes of mental retardation in females. The disease is almost always sporadic, with an estimated prevalence of 1:10,000/1:14,000. Girls with classic RTT show a normal neonatal period for the first months of life with subsequent developmental and neurological regression, characterized by head growth deceleration, loss of speech and purposeful use of hands, communication dysfunction, social withdrawal with autism, and stereotypical hand movements [91,92]. Most sporadic cases of classic RTT and around 50% of atypical cases are caused by *de novo* heterozygous mutations in the X-linked MECP2 gene, which stands for methyl-CpG-binding protein 2 [93]. Mutations are often linked to its three recognizable motifs, two of them (MBD and TRD [transcriptional repression domain]) involved in binding of the methylated DNA and of the Sin3A co-repressor, respectively, and the third one located at COOH end of the protein, similar to the fork head domain [94]. While the gene is broadly expressed [95], the protein is largely confined to the neurons [96].

Three mouse models have been described for MECP2 gene functional deletion. All of them give rise to profound behavioral phenotypes in the homozygous deleted [97,98] or mutated mice (MECP2 308/y; [99]), such as tremors, motor impairments, increased anxiety, seizures, and stereotypic hindlimb and forelimb clasping. Mice die after few weeks of postnatal life. Females heterozygotes develop similar symptoms, but later in life. All of these symptoms, as well as the timing of occurrence, are very similar to those exhibited by RTT patients. Mice deleted by a Cre-mediated excision of MECP2 gene, either in early neuronal development or in postnatal neurons, give rise to a similar phenotype, pointing to an effect on neuronal maintenance rather than development of brain functions. At a molecular level, analysis of the histone acetylation level in MECP2-mutated mice has identified hyperacetylation of histone H3 in brain tissues, thus supporting the notion that disruption of MECP2 function could lead to altered chromatin structure. Interestingly, the deletion of MBD2 gene gives a behavioral phenotype [100]. Since both MBD2 and MECP2 are expressed ubiquitously in humans and mice, the finding that disruption of either gene has a prominent effect on cerebral functions is somewhat surprising: functional redundancy of MBDs in the brain might be less than in other tissues and brain development may be exquisitely dependent upon DNA methylation-mediated silencing. A way to discriminate between these alternatives is represented by the identification of genes that are misregulated in the presence of dysfunctional MBDs. Direct binding of these proteins to specific gene promoters has been identified only in a few genes to date [101,102].

Expression profiling experiments have been recently done by using either autoptic brain tissues from RTT patients [103] or brain tissues derived by MECP2 knockout mice [104]. Both cases revealed no dramatic changes in transcription of specific genes. The results obtained suggest that, at least for MECP2, its deficiency leads to subtle gene expression changes in mutant brains. However, the role of MECP2 as transcriptional repressor and its importance for brain function has been confirmed recently by the identification of two putative targets: the transcriptional repressor of neuronal specific genes, Hairy2a [105], and the brain-derived growth factor (BDNF; [106–108]), which is thought to be essential for converting transient stimuli into long-term changes in brain activity. Interestingly, in this latter case, the over-expression of BDNF is caused by cortical depolarization that, on one hand phosphorylate MECP2, thus affecting its affinity for methylated DNA; on the other hand, it causes a limited hypomethylation of BDNF promoter III, thus boosting the effect. As already commented [108], these latest results may indicate that the MBD proteins, in addition to genome-wide functions, perform a number of gene-specific functions, thus revealing a functional plasticity and a biological role so far unsuspected.

FRAGILE X SYNDROME, ATR-X SYNDROME, AND FSHD: CPG METHYLATION, A USER'S GUIDE

A number of human syndromes share, among differential clinical presentations and molecular signs, the occurrence of differential DNA methylation that, in turn, might be involved as a primary cause of deregulation, or, rather, an effect of the causative gene dysfunction.

Fragile X Syndrome

Fragile X syndrome (FRA-X, OMIM #309520) is one of the most common causes of mental retardation, with an incidence of 1:4000 in males and 1:8000 in females. The syndrome is caused by mutations of the FMR-1 gene, located on the X chromosome. FRA-X was the first syndrome associated with trinucleotide expansion; in fact, the FMR-1 gene contains, at its 5UTR region, a CGG repeat that is highly polymorphic in normal individuals (between 7 and 60 repeats). Alleles with between 60 and 230 repeats are called permutations, whereas alleles with over 230 repeats are fully mutated [109]. The FMR-1 gene codifies for an RNA-binding protein, highly expressed in the brain. A step forward in the comprehension of its role in brain function has been made recently [110]: FMRP is a repressor of translation by regulating specific dendritic RNAs. FMRP is found in RNPs together with a non-translatable RNA, BC1.

What about the role of DNA methylation in FRA-X? The onset of the disease is connected with the CGG expansion at 5UTR; however, the expansion triggers cytosine methylation of the repeat and of flanking sequences, including FMR-1 gene promoter, thus inactivating the gene. Treatment of FRA-X cells with 5-aza-2-deoxy-citidine causes loss of methylation, reactivating FMR-1 transcription [111]. A synergistic effect of the combined treatment leading to histone hyperacetylation and DNA methylation has been demonstrated to reactivate FMR-1 transcription [112]. A detailed analysis of the histone modifications of the FMR-1 gene has recently been published [109], perhaps leading to the use of pharmaceutical treatment of this syndrome.

Alpha-Thalassemia Mental Retardation Syndrome (ATR-X)

ATR-X (OMIM #301040) is an X-linked mental retardation syndrome, characterized by severe mental retardation, reduced or absent speech, delayed developmental milestones, and congenital microcephaly. Alpha-thalassemia is a very frequent, but not constant, feature of the syndrome. Patients are mutated in the *ATRX* gene, located on Xq13 [113].

ATRX encodes for a putative ATP-dependent type II helicase of the SNF2 family. This protein binds the short arms of human acrocentric chromosomes and interacts with two heterochromatin proteins, mHP1a and EZH2. The ATRX protein belongs to an ATP-dependent chromatin remodeling complex, together with the transcription factor Daxx. This complex is localized to promyelocytic nuclear bodies [114].

The effects of ATRX mutations on the chromatin structure of rDNA arrays located at acrocentric chromosomes have been analyzed. Methylation differences have been noted: in ATR-X patients, rDNA genes were substantially unmethylated [115]. In addition to rDNA arrays, the analysis was extended to highly repetitive sequences spread in the genome. Abnormal methylation (hypomethylation) has been found at Y-chromosome-specific repeats, DYZ2, and subtelomeric repeats TelBam3.4. These features resemble those observed in ICF patients; however, in ATR-X, the amount of methylated cytosines is unaffected [115].

Facioscapulohumeral Muscular Dystrophy Syndrome (FSHD)

The autosomal-dominant myopathy, facioscapulohumeral muscular dystrophy syndrome (FSHD1, OMIM #158900), is the third most common myopathy, which progressively affects muscles of the face, shoulder, and upper arm. It has been associated with the contraction of the 3.3 kb repeat arrays D4Z4 on 4qter region [116]. Repeat numbers vary from 11 to 150 in normal populations, whereas it is greatly reduced (1 to 10) in FSHD patients. It was hypothesized that loss of D4Z4 repeats induces the region to adopt a more open chromatin structure, which bring to the up-regulation of target genes. The increased expression of three genes in FSHD, but not in unaffected muscle, has been reported. A protein complex drives the repression, recognizing a specific 27 bp element in the D4Z4 unit; this sequence has transcriptional repression activity *in vitro* [117]. However, a recent report links the marked hypomethylation of the contracted D4Z4 allele to overt disease [118]. This correlation is supported by the finding that FSHD patients with unaltered D4Z4 show hypomethylation of this repeat.

FUTURE DIRECTIONS

Epigenetic modifications and, more precisely, DNA methylation might be involved in all the conditions that fluctuate in severity as they progress and, in general, with pathologies in which the environment takes an important place. An example may be the differential methylation that occurs during aging: hypermethylation of specific oncosuppressor genes could be an important risk factor for the development of cancers.

DNA methylation can be affected by pharmaceutical treatments or dietary levels of methyl donor components, such as folic acid; folate supplementation of patients with hyperhomocysteinemia, secondary to chronic renal failure, directly affects gene expression, possibly as a function of the genomic DNA methylation level [79]. Given that in all the metabolic pathways involving methylation, the universal methyl donor is S-adenosylmethyonine, varying levels of any of the factors that feed into this methylation pathway could increase or decrease DNA or histone H3 methylation efficiency, affecting in both cases gene regulation [119]. There is evidence that epigenetic changes may be inherited, i.e., changes in diet could activate specific pathways, leaving an imprint that may be passed throughout generations. This has been postulated for diabetes in humans, as already reported [80,81], but firm data are reported only in the case of the coat-color gene agouti, in which changes from mottled to yellow are caused by the methylation state of an intragenic intracisternal A-particle element [60]. A diet supplemented with methyl donors affects the coat type of the offspring, thus establishing that a shift in coat color is correlated with the increase of DNA methylation [120]. A role for DNA methylation has been also proposed for behavioral disorders, such as in the reeler mouse, a mouse model for schizophrenia; the reelin gene shows affected promoter methylation and gene expression after a diet supplemented with L-methionine, whose effect is reversed by treatment with valproate [121]. This prompted researchers to analyze genomic methylation patterns from brain tissues of individuals affected by schizophrenia or bipolar disorders [122]. More ambitious is the project of the Human Epigenome Consortium [122,123], which aims to map genomic positions of distinct methylation variants and to build a dense genomic map, similar to those made for single nucleotide polymorphisms. Thus, the shifting in epigenetic profiles might be more consistent with such kinds of pathologies than the occurrence of susceptibility in genes, still in many cases unidentified, and may also provide the easiest route for pharmaceutical interventions.

REFERENCES

1. Scarano, E. et al., On methylation of DNA during development of the sea urchin embryo. *J. Mol. Biol.,* 14(2), 603–607, 1965.
2. Grippo, P., M. Iaccarino, E. Parisi, and E. Scarano, Methylation of DNA in developing sea urchin embryos. *J. Mol. Biol.,* 36, 195–208, 1968.
3. Riggs, A.D., X inactivation, differentiation, and DNA methylation. *Cytogenet. Cell Genet.,* 14(1), 9–25, 1975.

4. Bird, A.P., Gene number, noise reduction and biological complexity. *Trends Genet.*, 11(3), 94–100, 1995.
5. Yoder, J.A., C.P. Walsh, and T.H. Bestor, Cytosine methylation and the ecology of intragenomic parasites. *Trends Genet.*, 13(8), 335–340, 1997.
6. Lee, J.T., Molecular links between X-inactivation and autosomal imprinting: X-inactivation as a driving force for the evolution of imprinting? *Curr. Biol.*, 13(6), R242–254, 2003.
7. Li, E., T.H. Bestor, and R. Jaenisch, Targeted mutation of the DNA methyltransferase gene results in embryonic lethality. *Cell*, 69(6), 915–926, 1992
8. Okano, M. et al. DNA methyltransferases Dnmt3a and Dnmt3b are essential for *de novo* methylation and mammalian development. *Cell*, 99(3), 247–257, 1999.
9. Hendrich, B. and A. Bird, Identification and characterization of a family of mammalian methyl-CpG binding proteins. *Mol. Cell Biol.*, 18(11), 6538–6547, 1998.
10. Bird, A., DNA methylation patterns and epigenetic memory. *Genes Dev.*, 16(1), 6–21, 2002.
11. Hendrich, B. and S. Tweedie, The methyl-CpG binding domain and the evolving role of DNA methylation in animals. *Trends Genet.*, 19(5), 69–277, 2003.
12. Hendrich, B. and W. Bickmore, Human diseases with underlying defects in chromatin structure and modification. *Hum. Mol. Genet.*, 10(20), 2233–2242, 2001.
13. Bickmore, W.A. and S.M. van der Maarel, Perturbations of chromatin structure in human genetic disease: recent advances. *Hum. Mol. Genet.*, 12 Spec No 2: R207–213, 2003.
14. Cline, T.W. and B.J. Meyer, Vive la difference: males vs females in flies vs worms. *Annu. Rev. Genet.*, 30, 637–702, 1996.
15. Plath, K. et al., Xist RNA and the mechanism of X chromosome inactivation. *Annu. Rev. Genet.*, 36, 233–278, 2002.
16. Brockdorff, N. et al., The product of the mouse Xist gene is a 15 kb inactive X-specific transcript containing no conserved ORF and located in the nucleus. *Cell*, 71(3), 515–526, 2002.
17. Csankovszki, G., A. Nagy, and R. Jaenisch, Synergism of Xist RNA, DNA methylation, and histone hypoacetylation in maintaining X chromosome inactivation. *J. Cell Biol.*, 153(4), 773–784, 2001.
18. Takagi, N. and M. Sasaki, Preferential inactivation of the paternally derived X chromosome in the extraembryonic membranes of the mouse. *Nature*, 256(5519), 640–642, 1975.
19. Lee, J.T. et al. A 450 kb transgene displays properties of the mammalian X-inactivation center. *Cell*, 86(1), 83–94, 1996.
20. Clemson, C.M. et al. XIST RNA paints the inactive X chromosome at interphase: evidence for a novel RNA involved in nuclear/chromosome structure. *J. Cell Biol.*, 132(3), 259–275, 1996.
21. Avner, P. and E. Heard, X-chromosome inactivation: counting, choice and initiation. *Nat. Rev. Genet.*, 2(1), 59–67, 2001.
22. Wutz, A., T.P. Rasmussen, and R. Jaenisch, Chromosomal silencing and localization are mediated by different domains of Xist RNA. *Nat. Genet.*, 30(2), 167–174, 2002.
23. Lee, J.T., L.S. Davidow, and D. Warshawsky, Tsix, a gene antisense to Xist at the X-inactivation centre. *Nat. Genet.*, 21(4), 400–404, 1999.
24. Boumil, R.M. and J.T. Lee, Forty years of decoding the silence in X-chromosome inactivation. *Hum. Mol. Genet.*, 10(20): 2225–2232, 2001.
25. Lee, J.T. and N. Lu, Targeted mutagenesis of Tsix leads to nonrandom X inactivation. *Cell*, 99(1), 47–57, 1999.

26. Sado, T. et al., Regulation of imprinted X-chromosome inactivation in mice by Tsix. *Development,* 128(8), 1275–1286, 2001.

27. McDonald, L.E., C.A. Paterson, and G.F. Kay, Bisulfite genomic sequencing-derived methylation profile of the xist gene throughout early mouse development. *Genomics,* 54(3), 379–386, 1998.

28. Courtier, B., E. Heard, and P. Avner, Xce haplotypes show modified methylation in a region of the active X chromosome lying 3' to Xist. *Proc. Natl. Acad. Sci. USA,* 92(8), 3531–3535, 1995.

29. Prissette, M. et al., Methylation profiles of DXPas34 during the onset of X-inactivation. *Hum. Mol. Genet.,* 10(1), 31–38, 2001.

30. Lyon, M.F., X-chromosome inactivation. Pinpointing the centre. *Nature,* 379(6561), 116–117, 1996.

31. Bell, A.C., A.G. West, and G. Felsenfeld, The protein CTCF is required for the enhancer blocking activity of vertebrate insulators. *Cell,* 98(3), 387–396, 1998.

32. Chao, W. et al., CTcf. a candidate trans-acting factor for X-inactivation choice. *Science,* 295(5553), 345–337, 2002.

33. Wutz, A. and R. Jaenisch, A shift from reversible to irreversible X inactivation is triggered during ES cell differentiation. *Mol. Cell,* 5(4), 695–705, 2000.

34. Sado, T. et al., X inactivation in the mouse embryo deficient for Dnmt1: distinct effect of hypomethylation on imprinted and random X inactivation. *Dev. Biol.,* 225(2), 294–303, 2000.

35. Kaslow, D.C. and B.R. Migeon, DNA methylation stabilizes X chromosome inactivation in eutherians but not in marsupials: evidence for multistep maintenance of mammalian X dosage compensation. *Proc. Natl. Acad. Sci. USA,* 1987. 84(17), 6210–624.

36. Migeon, B.R. et al., Complete reactivation of X chromosomes from human chorionic villi with a switch to early DNA replication. *Proc. Natl. Acad. Sci. USA,* 83(7), 2182–2186, 1986.

37. Graves, J.A., 5-azacytidine-induced re-expression of alleles on the inactive X chromosome in a hybrid mouse cell line. *Ex. Cell Res.,* 141(1), 99–105, 1982.

38. Heard, E. et al., Methylation of histone H3 at Lys-9 is an early mark on the X chromosome during X inactivation. *Cell,* 107(6), 727–738, 2001.

39. Silva, J. et al. Establishment of histone h3 methylation on the inactive X chromosome requires transient recruitment of Eed-Enx1 polycomb group complexes. *Dev. Cell,* 4(4), 481–495, 2003.

40. Beard, C., E. Li, and R. Jaenisch, Loss of methylation activates Xist in somatic but not in embryonic cells. *Genes Dev.,* 9(19), 2325–2334, 1995.

41. Panning, B. and R. Jaenisch, DNA hypomethylation can activate Xist expression and silence X-linked genes. *Genes Dev.,* 10(16), 1991–1996, 2002.

42. Watanabe, D. et al., Stage- and cell-specific expression of Dnmt3a and Dnmt3b during embryogenesis. *Mech. Dev.,* 118(1-2), 187–190, 2002.

43. Hansen, R.S. et al., The DNMT3B DNA methyltransferase gene is mutated in the ICF immunodeficiency syndrome. *Proc. Natl. Acad. Sci. USA,* 96(25), 14412–14417, 1999.

44. Hansen, R.S. et al., Escape from gene silencing in ICF syndrome: evidence for advanced replication time as a major determinant. *Hum. Mol. Genet.,* 9(18), 2575–2587, 2000.

45. Hansen, R.S., X inactivation-specific methylation of LINE-1 elements by DNMT3B: implications for the Lyon repeat hypothesis. *Hum. Mol. Genet.,* 12(19), 2559–2567, 2003.

46. Lyon, M.F., X-chromosome inactivation: a repeat hypothesis. *Cytogenet. Cell Genet.*, 80(1-4), 133–137, 1998.
47. Kratzer, P.G. et al., Differences in the DNA of the inactive X chromosomes of fetal and extraembryonic tissues of mice. *Cell,* 33(1), 37–42, 1983.
48. Chapman, V. et al., Cell lineage-specific undermethylation of mouse repetitive DNA. *Nature,* 307(5948), 284–286, 1984.
49. Huynh, K.D. and J.T. Lee, Inheritance of a pre-inactivated paternal X chromosome in early mouse embryos. *Nature,* 426(6968), 857–862, 2003.
50. Surani, M.A., S.C. Barton, and M.L. Norris, Development of reconstituted mouse eggs suggests imprinting of the genome during gametogenesis. *Nature,* 308(5959), 548-550, 1984
51. McGrath, J. and D. Solter, Completion of mouse embryogenesis requires both the maternal and paternal genomes. *Cell,* 37(1), 179–183, 1984.
52. Nicholls, R.D. et al., Genetic imprinting suggested by maternal heterodisomy in nondeletion Prader-Willi syndrome. *Nature,* 42(6247), 281–285, 1989.
53. DeChiara, T.M., E.J. Robertson, and A. Efstratiadis, Parental imprinting of the mouse insulin-like growth factor II gene. *Cell,* 64(4), 849–859, 1991.
54. Reik, W. and J. Walter, Genomic imprinting: parental influence on the genome. *Nat. Rev. Genet.,* 2(1), 21–32, 2001.
55. Walter, J. and M. Paulsen, The potential role of gene duplications in the evolution of imprinting mechanisms. *Hum. Mol. Genet.,* 12 Spec No 2, R215–220, 2003.
56. Kaneko-Ishino, T., T. Kohda, and F. Ishino, The regulation and biological significance of genomic imprinting in mammals. *J. Biochem.* (Tokyo), 133(6), 699–711, 2003.
57. Li, E., Chromatin modification and epigenetic reprogramming in mammalian development. *Nat. Rev. Genet.,* 3(9), 662–673, 2002.
58. Kitsberg, D. et al., Allele-specific replication timing of imprinted gene regions. *Nature,* 364(6436), 459–463, 1993., G. Gyapay, and J. Jami, Imprinted chromosomal regions of the human genome display sex-specific meiotic recombination frequencies. *Curr. Biol.,* 5(9), 1030–1035, 1993.
60. Jaenisch, R. and A. Bird, Epigenetic regulation of gene expression: how the genome integrates intrinsic and environmental signals. *Nat. Genet.,* 33 Suppl: 245–254, 2003.
61. Geuns, E. et al., Methylation imprints of the imprint control region of the SNRPN-gene in human gametes and preimplantation embryos. *Hum. Mol. Genet.,* 12(22), 2873–2879, 2003.
62. Dorer, D.R. and S. Henikoff, Expansions of transgene repeats cause heterochromatin formation and gene silencing in Drosophila. *Cell,* 77(7), 993–1002, 2004.
63. Loukinov, D.I. et al., BORIS, a novel male germ-line-specific protein associated with epigenetic reprogramming events, shares the same 11-zinc-finger domain with CTCF, the insulator protein involved in reading imprinting marks in the soma. *Proc. Natl. Acad. Sci. USA,* 99(10), 6806–6811, 2002.
64. Feinberg, A.P., M. Oshimura, and J.C. Barrett, Epigenetic mechanisms in human disease. *Cancer Res.,* 62(22), 6784–6787, 2002.
65. Nicholls, R.D., The impact of genomic imprinting for neurobehavioral and developmental disorders. *J. Clin. Invest.,* 105(4), 413–418, 2002.
66. Falls, J.G. et al., Genomic imprinting: implications for human disease. *Am. J. Pathol.,* 154(3), 635–647, 1999.
67. Clayton-Smith, J. and L. Laan, Angelman syndrome: a review of the clinical and genetic aspects. *J. Med. Genet.,* 40(2), 87–95, 2003.

68. Nicholls, R.D. and J.L. Knepper, Genome organization, function, and imprinting in Prader-Willi and Angelman syndromes. *Annu. Rev. Genomics Hum. Genet.*, 2, 153–175, 2001.

69. Brannan, C.I. and M.S. Bartolomei, Mechanisms of genomic imprinting. *Curr. Opin. Genet. Dev.*, 9(2): 164–170, 1999.

70. Buiting, K. et al., Inherited microdeletions in the Angelman and Prader-Willi syndromes define an imprinting centre on human chromosome 15. *Nat. Genet.*, 9(4): 395–400, 1995.

71. Bittel, D.C. et al., Microarray analysis of gene/transcript expression in Prader-Willi syndrome: deletion versus UPD. *J. Med. Genet.*, 40(8), 568–574, 2003.

72. Albrecht, U. et al., Imprinted expression of the murine Angelman syndrome gene, Ube3a, in hippocampal and Purkinje neurons. *Nat. Genet.*, 17(1), 75-78, 1997.

73. Runte, M. et al., The IC-SNURF-SNRPN transcript serves as a host for multiple small nucleolar RNA species and as an antisense RNA for UBE3A. *Hum. Mol. Genet.*, 10(23), 2687–2700, 2001.

74. Gallagher, R.C. et al., Evidence for the role of PWCR1/HBII-85 C/D box small nucleolar RNAs in Prader-Willi syndrome. *Am. J. Hum. Genet.*, 71(3), 669–678, 2002.

75. Weksberg, R. et al., Beckwith-Wiedemann syndrome demonstrates a role for epigenetic control of normal development. *Hum. Mol. Genet.*, 12 Spec No 1: R61–68, 2003.

76. Maher, E.R. and W. Reik, Beckwith-Wiedemann syndrome: imprinting in clusters revisited. *J. Clin. Invest.*, 105(3), 247–252, 2000.

77. Neyroud, N. et al., A novel mutation in the potassium channel gene KVLQT1 causes the Jervell and Lange-Nielsen cardioauditory syndrome. *Nat. Genet.*, 15(2), 186–189, 1997.

78. Kanduri, C. et al., A differentially methylated imprinting control region within the Kcnq1 locus harbors a methylation-sensitive chromatin insulator. *J. Biol. Chem.*, 277(20), 18106–18110, 2000.

79. Ingrosso, D. et al., Folate treatment and unbalanced methylation and changes of allelic expression induced by hyperhomocysteinaemia in patients with uraemia. *Lancet*, 361(9370), 1693–1693, 2003.

80. Kaati, G., L.O. Bygren, and S. Edvinsson, Cardiovascular and diabetes mortality determined by nutrition during parents' and grandparents' slow growth period. *Eur. J. Hum. Genet.*, 10(11), 682–688, 2002.

81. Pembrey, M.E., Time to take epigenetic inheritance seriously. *Eur. J. Hum. Genet.*, 10(11), 669–671, 2002.

82. Wijmenga, C. et al., Localization of the ICF syndrome to chromosome 20 by homozygosity mapping. *Am. J. Hum. Genet.*, 63(3), 803–809, 1998.

83. Xu, G.L. et al., Chromosome instability and immunodeficiency syndrome caused by mutations in a DNA methyltransferase gene. *Nature*, 402(6758), 187–191, 1999.

84. Ehrlich, M., The ICF syndrome, a DNA methyltransferase 3B deficiency and immunodeficiency disease. *Clin. Immunol.*, 109(1), 17–28, 2003.

85. Jeanpierre, M. et al., An embryonic-like methylation pattern of classical satellite DNA is observed in ICF syndrome. *Hum. Mol. Genet.*, 2(6), 731–735, 1993.

86. Miniou, P. et al., Abnormal methylation pattern in constitutive and facultative (X inactive chromosome) heterochromatin of ICF patients. *Hum. Mol. Genet.*, 3(12), 2093–2102, 1994

87. Bourc'his, D., et al., Abnormal methylation does not prevent X inactivation in ICF patients. *Cytogenet. Cell Genet.*, 84(3-4), 245–252, 1999.

88. Kondo, T. et al., Whole-genome methylation scan in ICF syndrome: hypomethylation of non-satellite DNA repeats D4Z4 and NBL2. *Hum. Mol. Genet.,* 9(4), 597–604, 2000.

89. Matarazzo, M.R., et al., Allelic inactivation of the pseudoautosomal gene SYBL1 is controlled by epigenetic mechanisms common to the X and Y chromosomes. *Hum. Mol. Genet.,* 11(25), 3191–3198, 2004.

90. Ehrlich, M. et al., DNA methyltransferase 3B mutations linked to the ICF syndrome cause dysregulation of lymphogenesis genes. *Hum. Mol. Genet.,* 10(25), 2917–2931, 2001.

91. Rett, A., Über ein eigenartiges hirnatrophisches Syndrom bei Hyperammonemie im Kindesalter. *Wien. Med. Wochenschr.,* 116, 723–728, 1966.

92. Hagberg, B., J. Aicardi, K. Dias, and O. Ramos, A progressive syndrome of autism, dementia, ataxia, and loss of purposeful hand use in girls: Rett's syndrome: report of 35 cases. *Ann. Neurol.,* 4: 471–479, 1983.

93. Amir, R.E. et al., Rett syndrome is caused by mutations in X-linked MECP2, encoding methyl-CpG-binding protein 2. *Nat. Genet.,* 23(2), 185–188, 1999.

94. Vacca, M. et al., Mutation analysis of the MECP2 gene in British and Italian Rett syndrome females. *J. Mol. Med.,* 78(11), 648–655, 2001a.

95. D'Esposito, M., et al., Isolation, physical mapping, and northern analysis of the X-linked human gene encoding methyl CpG-binding protein, MECP2. *Mamm. Genome,* 7(7), 533–535, 1996.

96. LaSalle, J.M. et al,. Quantitative localization of heterogeneous methyl-CpG-binding protein 2 (MeCP2) expression phenotypes in normal and Rett syndrome brain by laser scanning cytometry. *Hum. Mol. Genet.,* 10(17), 1729–1740, 2001.

97. Chen, R.Z. et al., Deficiency of methyl-CpG binding protein-2 in CNS neurons results in a Rett-like phenotype in mice. *Nat. Genet.,* 27(3), 327–331, 2001.

98. Guy, J., et al., A mouse Mecp2-null mutation causes neurological symptoms that mimic Rett syndrome. *Nat. Genet.,* 27(3), 322–326, 2001.

99. Shahbazian, M., et al., Mice with truncated MeCP2 recapitulate many Rett syndrome features and display hyperacetylation of histone H3. *Neuron,* 35(2),: 243–254, 2002.

100. Hendrich, B. et al., Closely related proteins MBD2 and MBD3 play distinctive but interacting roles in mouse development. *Genes Dev.,* 15(6), 710–723, 2001.

101. Fournier, C. et al., Allele-specific histone lysine methylation marks regulatory regions at imprinted mouse genes. *Embo. J.,* 21(23), 6560–6570, 2002.

102. Ballestar, E. et al., Methyl-CpG binding proteins identify novel sites of epigenetic inactivation in human cancer. *Embo. J.,* 22(23), 6335–6345, 2003.

103. Colantuoni, C. et al., Gene expression profiling in postmortem Rett Syndrome brain: differential gene expression and patient classification. *Neurobiol. Dis.,* 8(5), 847–865, 2001.

104. Tudor, M. et al., Transcriptional profiling of a mouse model for Rett syndrome reveals subtle transcriptional changes in the brain. *Proc. Natl. Acad. Sci. USA,* 99(24), 15536–15541, 2002.

105. Stancheva, I. et al., A mutant form of MeCP2 protein associated with human Rett syndrome cannot be displaced from methylated DNA by notch in Xenopus embryos. *Mol. Cell,* 12(2), 425–435, 2003

106. Martinowich, K. et al., DNA methylation-related chromatin remodeling in activity-dependent BDNF gene regulation. *Science,* 302(5646), 890–893, 2003.

107. Chen, W.G. et al., Derepression of BDNF transcription involves calcium-dependent phosphorylation of MeCP2. *Science,* 302(5646), 885–889, 2003.

108. Klose, R. and A. Bird, Molecular biology. MeCP2 repression goes nonglobal. *Science,* 302(5646), 793–795, 2003.

109. Coffee, B. et al., Histone modifications depict an aberrantly heterochromatinized FMR1 gene in fragile x syndrome. *Am. J. Hum. Genet.,* 71(4), 923–932, 2002.

110. Zalfa, F. et al., The fragile X syndrome protein FMRP associates with BC1 RNA and regulates the translation of specific mRNAs at synapses. *Cell,* 112(3), 317–327, 2003.

111. Chiurazzi, P. et al., In vitro reactivation of the FMR1 gene involved in fragile X syndrome. *Hum. Mol. Genet.,* 7(1): 109–113, 1998.

112. Chiurazzi, P. et al., Synergistic effect of histone hyperacetylation and DNA demethylation in the reactivation of the FMR1 gene. *Hum. Mol. Genet.,* 8(12), 2317–2323, 1999.

113. Gibbons, R.J. et al., Mutations in a putative global transcriptional regulator cause X-linked mental retardation with alpha-thalassemia (ATR-X syndrome). *Cell,* 80(6), 837–845, 1995.

114. Xue, Y. et al., The ATRX syndrome protein forms a chromatin-remodeling complex with Daxx and localizes in promyelocytic leukemia nuclear bodies. *Proc. Natl. Acad. Sci. USA,* 100(19), 10635–10640, 2003.

115. Gibbons, R.J. et al., Mutations in ATRX, encoding a SWI/SNF-like protein, cause diverse changes in the pattern of DNA methylation. *Nat. Genet.,* 24(4), 368–371, 2000.

116. Wijmenga, C. et al., Chromosome 4q DNA rearrangements associated with facioscapulohumeral muscular dystrophy. *Nat. Genet.,* 2(1), 26–3, 1992.

117. Gabellini, D., M.R. Green, and R. Tupler, Inappropriate gene activation in FSHD: a repressor complex binds a chromosomal repeat deleted in dystrophic muscle. *Cell,* 110(3), 339–348, 2002.

118. van Overveld, P.G. et al., Hypomethylation of D4Z4 in 4q-linked and non-4q-linked facioscapulohumeral muscular dystrophy. *Nat. Genet.,* 35(4), 315–317, 2003.

119. Van den Veyver, I.B., Genetic effects of methylation diets. *Annu. Rev. Nutr.,* 22, 255–282, 2002.

120. Cooney, C.A., A.A. Dave, and G.L. Wolff, Maternal methyl supplements in mice affect epigenetic variation and DNA methylation of offspring. *J. Nutr.,* 132(8 Suppl), 2393S–2400S, 2002.

121. Tremolizzo, L. et al., An epigenetic mouse model for molecular and behavioral neuropathologies related to schizophrenia vulnerability. *Proc. Natl. Acad. Sci. USA,* 99(26), 17095–17100, 2002.

122. Dennis, C., Epigenetics and disease: altered states. *Nature,* 421(6924), 686–688, 2003.

123. Beck, S., A. Olek, and J. Walter, From genomics to epigenomics: a loftier view of life. *Nat. Biotechnol.,* 17(12), 114, 1999.

4 Studying Mammalian DNA Methylation: Bisulfite Modification

Susan J. Clark

CONTENTS

ACKNOWLEDGMENTS

I wish to thank Wenjia Qu and Rebecca Hinschelwood for help with figures and Peter Molloy for reading the manuscript. The methylation research in Dr. Clark's lab is funded by National Health and Medical Research Council grants 293810 and 202907.

ABSTRACT

This chapter summarizes the current user-friendly laboratory protocol for the study of methylation using bisulfite conversion. The process of bisulfite treatment exploits the different sensitivity of cytosine and 5-methylcytosine (5-MeC) to deamination by bisulfite under acidic conditions, in which cytosine undergoes conversion to uracil while 5-MeC remains unreactive. Methylation analysis after bisulfite conversion has become the most widely used method to detect 5-MeC in DNA and provides a reliable way of detecting any methylated cytosine even at single molecule resolution. However, the method by which the converted DNA is analyzed can influence the interpretation of the methylation status of the DNA. In this chapter, we discuss the different approaches that can be used for DNA methylation analysis depending on the degree of specificity and detail required. We suggest ways to reduce amplification of unconverted DNA, and compare results using direct polymerase chain reaction (PCR) sequencing versus clonal analysis to determine the methylation pattern of a sequence.

INTRODUCTION

The detection of 5-MeC using bisulfite conversion was first reported by Frommer et al.[1] and enabled by Clark et al.[2] Since then, the validity of bisulfite sequencing has been confirmed to the point where methods based around bisulfite conversion account for the majority of new data on DNA methylation. Methylation analysis using bisulfite sequencing has been used to study the change in methylation patterns in specific sequences during early development, tissue specificity of methylation, and aberrant methylation in cancer. The most significant result to come from this early sequencing work was the finding that not all CpG sites are methylated equally and that the degree of heterogeneity at any one site in a sequence suggests that the interplay between *de novo* methylation and maintenance methylation is more dynamic than first predicted.[3,4] Thus bisulfite sequencing has allowed the complexity of methylation biology to be appreciated and has opened the door to explore more questions about the process of DNA methylation, both in normal and cancer cells.

PRINCIPLE OF BISULFITE CONVERSION

The key to determining methylated cytosines is based on the selective chemical reaction of sodium bisulfite with cytosine versus methylated cytosine residues. The reaction is highly single strand specific and, therefore, it is important to ensure that the genomic DNA is fully denatured prior to the bisulfite reaction. Cytosine can form adducts across the 5-6 bond with the bisulfite ion. The mechanism of this reaction was initially reported in the early 1970s.[5] The deamination of cytosine by sodium bisulfite, as summarized in Figure 4.1, involves the following steps: (1) *Sulfonation*: addition of bisulfite to the 5-6 double bond of cytosine, (2) *Hydrolytic deamination*: deamination of the resulting cytosine–bisulfite derivative to give a uracil–bisulfite derivative, and (3) *Alkali desulfonation*: removal of the sulfonate group by a subsequent alkali treatment to give uracil. The extent of sulfonation formation is controlled by pH, bisulfite concentration, and temperature.[2,6,7] The forward sulfonation reaction is favored by low pH and the reverse reaction by high pH. In the second step of the reaction, cytosine–SO_3 undergoes hydrolytic deamination to give uracil–SO_3. This step is catalyzed by basic substances, such as sulfite, bisulfite, and acetate anions.[5] Since sulfonation is favored by acidic pH, the reversible sulfonation reaction and the subsequent irreversible deamination step are both carried out at pH 5 in high bisulfite concentration. Bisulfite can oxidize automatically with oxygen, and so a free radical scavenger is included in the reaction to minimize oxidative degradation. The third step of the reaction involves alkali treatment to change the pH and remove the bisulfite adduct.

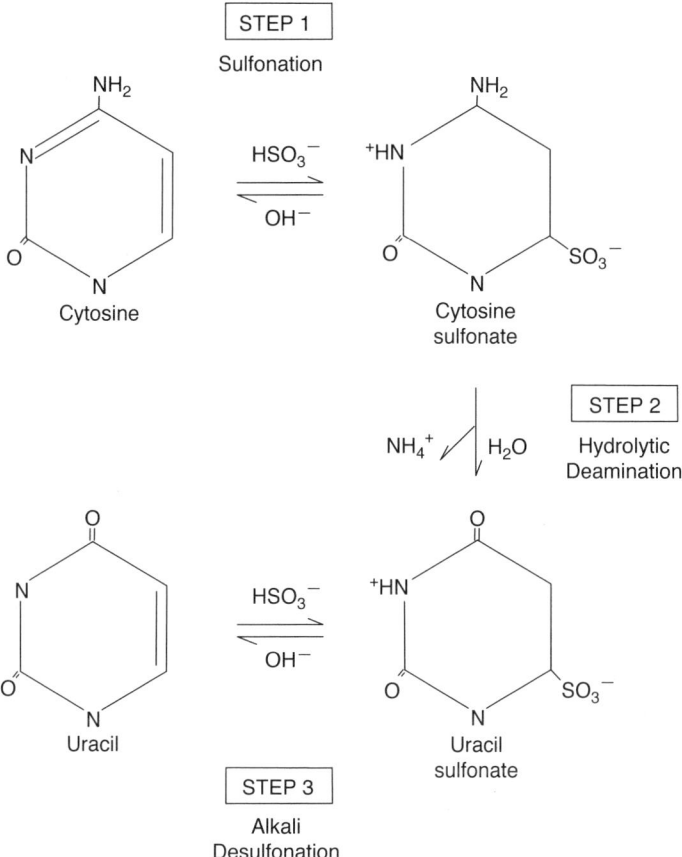

FIGURE 4.1 Schematic diagram of the bisulfite conversion reaction.[2]

The important conditions for a "successful" bisulfite conversion are (1) clean and fully denatured DNA; (2) freshly prepared bisulfite; (3) low pH for the deamination reaction; (4) high pH for the desulfonation reaction; and (5) hydroquinone to reduce bisulfite oxidation. The time of incubation varies with the amount and source of DNA to be treated. As a general rule, a small amount of low-quality DNA, such as from bodily fluids, requires only short incubation times (4 to 8 h), whereas high concentrations of DNA, such as from cell lines or tissues, can require up to 16 h of incubation to ensure complete conversion.[8] The "standard" bisulfite protocol that can be used for all DNA templates, including DNA extracted from bodily fluids, is described below.

"STANDARD" BISULFITE PROTOCOL

1. The source of DNA can be from cell lines, tissues, or bodily fluids. It is preferable to use DNA that is free of proteins by prior digestion with

proteinase K. For limited cell numbers (10 to 1000 cells), we resuspend the fresh, frozen, or fixed cells in 18 μl proteinase K/SDS buffer (2 μg *E.coli* tRNA, 1% SDS, 280 μg/ml proteinase K) and incubate at 37°C for 30 min prior to the addition of sodium hydroxide.

2. For large molecular weight DNA, it is preferable either to digest with an appropriate restriction enzyme (avoiding digestion of the target region), or to shear by passing the DNA resuspended in proteinase K/SDS buffer, 5 times in through a 26 gauge needle, followed by incubation at 37°C for 30 min (reducing the size of the DNA aids in ensuring full denaturation).

3. Add 2.2 μl freshly prepared 3M NaOH to the DNA sample in 18 μl; vortex and centrifuge briefly.

4. Incubate at 37°C for 15 min.

5. Incubate at 90°C for 2 min, place on ice, and centrifuge briefly.

6. Add 208 μl saturated sodium metabisulfite pH 5.0 (BDH). (Saturated sodium metabisulfite is prepared by adding 7.6 g $Na_2S_2O_5$ to 15 ml water and adjusting the pH to 5.0 by adding 464 μl freshly made 10M NaOH.)

7. Add 12 μl 10 mM hydroquinone; vortex and centrifuge briefly.

8. Overlay the samples with 200 μl mineral oil to prevent evaporation and limit oxidation.

9. Incubate for 4 to 16 h at 55°C in a water bath with a lid.

10. Remove the mineral oil and discard.

11. Add 1 ml Promega's Wizard® DNA Clean-Up resin and desalt according to manufacturer's instructions.

12. Add 50 μl water to the minicolumn and leave at room temperature for 5 min.

13. Spin for 20 sec to collect the eluant and discard the minicolumns.

14. Add 5.5 μl 3M NaOH and incubate at 37°C for 15 min.

15. Centrifuge briefly.

16. Add 1 μl tRNA (10 mg/ml).

17. Add 33.3 μl 5M $NH_4OAcetate$ pH 7.0.

18. Add 330 μl cold (–20°C) 100% ethanol and mix well.

19. Leave at –20°C overnight.

20. Centrifuge at 14,000 rpm for 15 min at 4°C.

21. Resuspend the DNA pellet in 10 μl H_2O at room temperature for ~2 h with occasional vortexing.

22. Store at –20°C in freezer.

EFFICIENCY OF BISULFITE CONVERSION

After bisulfite treatment, the unmethylated cytosines will be converted to thymine residues, and the methylated cytosines will remain intact. The conversion of cytosine to uracil by bisulfite is remarkably selective and efficient when carried out under the "standard" protocol. The rate of chemical conversion has been estimated to be in the order of 99.5 to 99.7%[7] of all cytosines, but is more often 95 to 98% due to varying DNA quality. For most applications this rate of conversion is more than sufficient, and the low level of unconverted cytosines is detectable only by a detailed analysis of cloned fragments. However, if the rate of conversion is less efficient or variable within a given DNA sequence, then an increased level of cytosines remain in the sequenced molecule, usually at non-CpG positions and in contiguous DNA regions or in blocks. Blocked regions represent areas of DNA that were nonreactive to bisulfite deamination due to interference from bound proteins,[9] insufficient denaturation, or rapid renaturation. Blocked regions are commonly encountered in repeated sequences and in CpG-rich DNA with secondary structure.[10] It is, therefore, important to perform multiple independent bisulfite reactions on protein-free DNA if regions of apparent nonconversion are observed. Often non-CpG methylation has been confused with a lack of bisulfite conversion.

DETECTION OF METHYLATION BY PCR AMPLIFICATION

The target region to be analyzed for methylation is amplified from the bisulfite-converted DNA using specific PCR primers. The type of primers chosen will greatly influence the methylation result obtained (summarized in Table 4.1). Extensively methylated DNA regions will be preferentially amplified by using methylation-specific PCR (MSP) primers.[11] Conversely, hypomethylated DNA regions will be preferentially amplified by using unmethylated-DNA-specific PCR (USP) primers. Using these primer sets, a simple yes or no answer for methylation is obtained, and complete methylation, or complete absence of methylation across the region, is assumed. However, in most studies, the degree of methylation is not known and, therefore, use of nonselective PCR (NSP) primers is required.[2] NSP primers are

TABLE 4.1
Primer Selection Depends on Methylation Profile

Methylation Profile	Fully Methylated	Fully Unmethylated	Heterogeneous Methylation	Differentially Methylated
Examples of gene targets	CpG islands promoters in tissue culture	CpG island promoters in normal cells	CpG island promoters in cancer	Imprinted and X-linked genes
Primer selection	MSP	USP	NSP	NSP
Primer length	20–25	20–25	25–30	25–30
% cytosine in primer for conversion	25–30%	25–30%	25–30% end in 3′C	25–30% end in 3′C
Primer CpG content	3–4 CpGs	3–4 TpG	0%	0%
Position of CpGs	3′ end is a C of a CpG	3′ end is a T of a TpG		

designed with no CpG in the priming region or mismatches to CpGs and so theoretically can amplify methylated and unmethylated DNA equally.[12] This feature is important when the DNA is heterogeneously methylated in the target region, or when quantitation of methylation at a given site is required. In addition, if there is differential methylation in the target region, such as for imprinted or X-linked genes, it is important to use NSP primers so that both copies of the gene will be represented. In this case, it is also important to optimize the PCR conditions such that both methylated and unmethylated DNA amplify in similar proportions.[13] This may not necessarily be the case, especially in CpG-rich regions, as the melting temperature, polymerization, or copying efficiency of the template (C-rich), versus the unmethylated template (T-rich), will vary considerably.

Whatever priming sets are chosen, it is essential to ensure amplification and/or detection of only fully converted DNA. With this in mind, it is important to design primers to a region that is as rich as possible in cytosine resides. However, even with these precautions MSP has a tendency to amplify regions of unconverted DNA. This is because the cytosine residue of the CpG is preferably placed at the 3 end of the primer to ensure selective amplification of methylated DNA; however, if a small proportion of the region is unconverted in the bisulfite-treated DNA reaction, these molecules may also be selectively amplified. Because of this problem of selective amplification, the number of false positives with MSP is often high. To reduce the number of false methylation positives scored, we previously have recommended using a hybridization probe for full conversion, thereby enabling only the methylated and converted amplified DNA to be quantified.[14]

Another approach to avoid false positives is to reduce the temperature of denaturation in the PCR amplification cycling. Unconverted DNA has a higher denaturation temperature than fully converted DNA and often amplifies more efficiently than converted DNA. Figure 4.2 shows an example in which a dilution series from 1 to 10^6 copies of bisulfite-converted *GSTP1* (glutathione S-transferase) DNA was spiked into a small amount of bisulfite-unconverted *GSTP1* DNA, and these mixtures were used as templates for MSP. At a denaturation temperature of 95°C, only the unconverted *GSTP1* DNA was amplified, even when 10^6 copies of converted DNA was in the amplification reaction (Figures 4.2A and B). However, if the denaturation temperature was reduced to 80°C in the PCR cycling reaction, the unconverted DNA, which previously masked the presence of methylated DNA, was no longer amplified (Figures 4.2C and D). Moreover, the lower denaturation temperature permitted the spiked, converted, and methylated DNA copies to be amplified in proportion to their input level. Ablation of the unconverted DNA in the PCR is due to the fact that the C-rich, unconverted amplicon has a higher melting temperature and at 80°C remains double stranded and, therefore, no longer provides a template in the PCR amplification reaction. In contrast, the converted DNA is relatively T-rich and has a lower melting temperature and denatures at 80°C, providing the only single-strand template for amplification in the reaction mix. Using a combination of differential temperature and a conversion-specific probe with real-time PCR allows for an accurate high-throughput screen of methylated molecules.

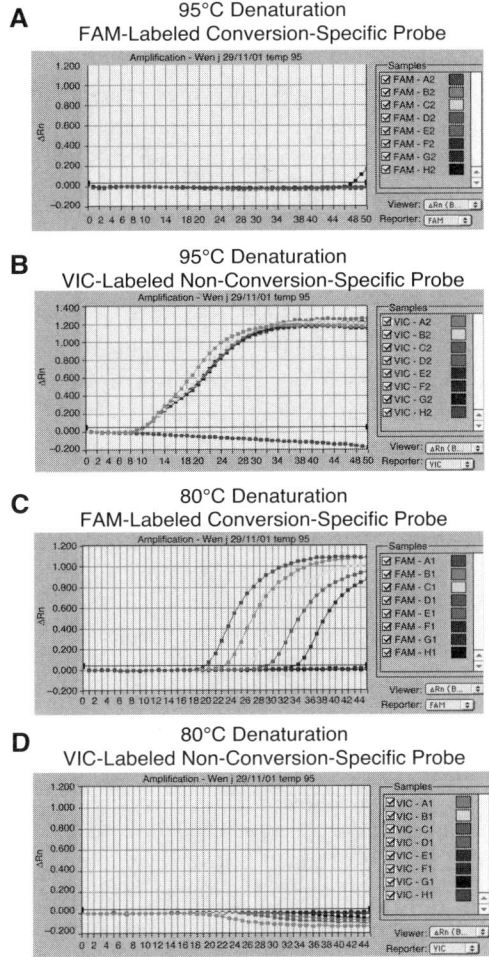

FIGURE 4.2 (See color insert following page 114) Effect of the denaturation temperature in the PCR cycling reaction on eradicating detection of unconverted DNA. 0, 10, 10^2, 10^3, 10^4, 10^5, and 10^6 copies of a fully converted and methylated plasmid clone containing *GSTP1* CpG island (Profiles A–G respectively) were spiked into unconverted DNA. Profile H shows the no DNA control. The *GSTP1* promoter region was amplified using MSP primers (MSP-81: 5-TTTTCGCGATGTTTCGGC and MSP-82: 5-GCCGCGCAACTAACCGA) and the amplicons detected by real-time PCR using either a FAM-labeled probe to converted *GSTP1* DNA (5-FAM-TTGCGTATATTTCGTTGCGGTTTTTTTTTT-TAMRA) or a VIC-labeled probe to unconverted *GSTP1* DNA (5-VIC-CTGTCTGTTTACTCCCTAGGCCC-TAMRA), as described in Rand et al., 2002.[14] (**A**) Denaturation temperature of 95°C. Cycling conditions: 95°C × 2 min, (95°C × 15 sec, 60°C × 1 min) × 50 cycles. Detection probe: FAM-labeled probe to converted *GSTP1* DNA. (**B**) Denaturation temperature of 95°C. Cycling conditions: 95°C × 2 min, (95°C × 15 sec, 60°C × 1 min) × 50 cycles. VIC-labeled probe to unconverted *GSTP1* DNA. (**C**) Denaturation temperature of 80°C. Cycling conditions: 95°C × 2 min, (95°C × 15 sec, 60°C × 1 min) × 5 cycles; (80°C × 15 sec, 60°C × 1 min) × 45 cycles. FAM-labeled probe to converted *GSTP1* DNA. (**D**) Denaturation temperature of 80°C. Cycling conditions: 95°C × 2 min, (95°C × 15 sec, 60°C × 1 min) × 5 cycles; (80°C × 15 sec, 60°C × 1 min) × 45 cycles. VIC-labeled probe to unconverted *GSTP1* DNA. At a denaturation temperature of 95°C, only unconverted DNA was amplified. However, if the denaturation temperature was reduced to 80°C, only the converted DNA was amplified and the amplification was quantitative.

DETECTION OF METHYLATION BY
DNA SEQUENCING

If methylation analysis of only single CpG sites is required, then the amplified bisulfite-treated product can be assayed using informative restriction enzymes, such as *Taq*1 (TCGA) or *Bst*U1 (CGCG), or by using a SNuPE-based assay.[15] However, methylation sequencing is essential if contiguous CpG sites are to be analyzed in a defined region. To ensure a representative DNA methylation profile, we recommend that each bisulfite DNA be amplified in triplicate and then the PCR products pooled. The pooled PCR products can be directly sequenced using the forward and reverse amplification primers as sequencing primers. Figure 4.3A shows a direct sequencing profile of the ICAM-1 gene. The relative peak heights of the cytosine to thymine peaks at each site can be used to semiquantitate the methylation levels (Figure 4.3B).

For a more detailed methylation profile, in particular, if understanding methylation heterogeneity of a region is important, then the pooled PCR products can be cloned and individually sequenced. Pooling is again important to ensure an accurate profiling of the genomic DNA, as often single PCR products may contain only a limited number of molecules. Figure 4.3C shows the difference in the methylation profile by direct PCR sequence analysis versus clonal sequence analysis. The direct sequence profile reflects the density of methylation in the population, whereas the clonal sequencing portrays the degree of heterogeneity in the cells. Moreover, if methylation levels are low (less than 25% at any one site), then direct sequencing will not be sufficiently sensitive to be reliable for methyaltion detection, and cloning is required. To ensure a representation of the molecules in the PCR amplifications, at least 10 independent clones are required, but often more are needed if the methylation pattern is particularly heterogeneous.

Even though cloning and sequencing may be laborious and expensive, the insight into methylation biology that is obtained is invaluable to understanding the process and the role of methylation in the cell. The human and mouse genome have been sequenced for the four bases, but bisulfite methylation sequencing is now required to provide yet another dimension of information that is potentially provided by selective methylation of cytosine residues in different cell types, and at different stages of development. The implication of methylation changes, both in disease states and during aging, have exciting diagnostic and therapeutic potential.

CONCLUSION

Bisulfite conversion permits differential determination of methylated cytosines from unmethyalted cytosines. However, the method of analysis of the bisulfite-treated DNA is critical, as it influences the interpretation of the methylation pattern in the genome and can result in misleading conclusions. Before choosing between the different methods available for analysis, the expected methylation profile of the DNA region should be taken into account. To determine the methylation profile of a given target region, we recommend that the level of heterogeneity of methylation of that region first be assessed using bisulfite sequencing. Understanding the methylation profile is especially important prior to the design of primers for a more routine MSP

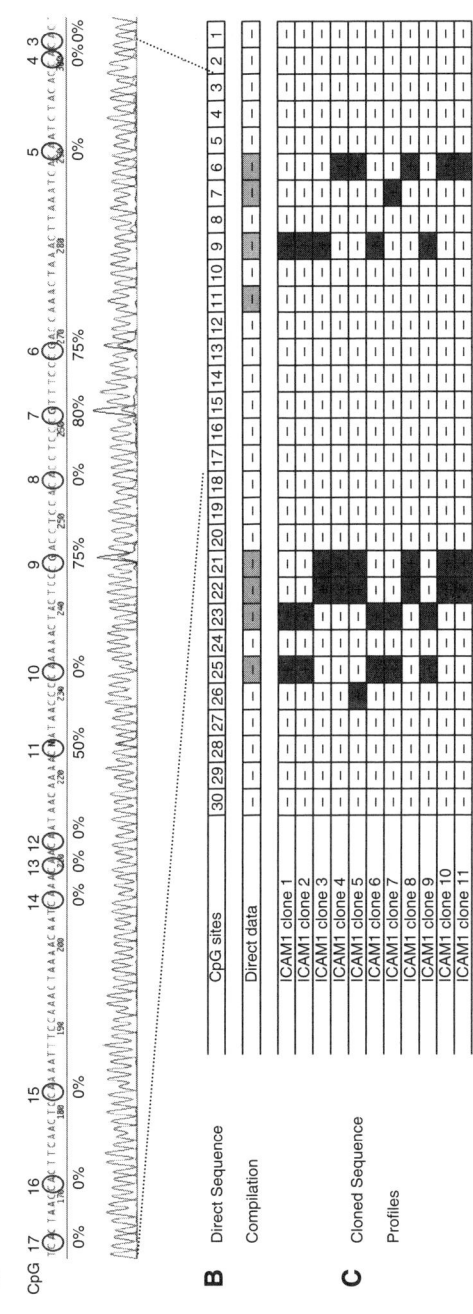

FIGURE 4.3   (See color insert) Bisulfite sequencing of the ICAM-1 gene. (**A**) An example of the direct sequencing profile obtained using the reverse primer of the PCR amplification in the Dye Terminator cycle sequencing kit with AmpliTaq DNA polymerase and the automated 373A Sequencer sequencing software Model 3700 version 3.3 (Applied Biosystems). The degree of methylation is calculated by comparing the peak height of the cytosine residues with the peak of the thymine residues. (**B**) A compilation of the direct sequencing profile from CpG sites 1 to 30 within the promoter of the ICAM-1 gene. The shaded areas represent sites of methylation and the unshaded areas represent unmethylated sites. (**C**) Sequencing results from individual clones of the PCR product that was directly sequenced. A number of slightly different methylation patterns are represented in the clones, but the average methylation at each site equates with the direct PCR sequencing profile. However, if a site is methylated in only a limited number of clones, such as CpG site 26, this degree of methylation is not detected by direct PCR sequencing. In addition, CpG site 11 in the direct PCR sequence shows 50% methylation, but this site is not represented in the clones, suggesting that more clones are required to fully represent the methylation state of the target region.

or real-time analysis of the DNA samples. Methylation sequence information allows primers to be designed successfully by taking into account the methylation propensity of individual CpG sites within the region. The tools available for methylation detection have become more sophisticated and more informative over the last decade, but care and thoughtful design of the experiment and analysis methods chosen are critical.

REFERENCES

1. Frommer, M., McDonald, L. E., Millar, D.S., Collis, C. M., Watt, F., Grigg, G. W., Molloy, P. L., and Paul, C. L., A genomic sequencing protocol that yields a positive display of 5-methylcytoine residues in individual DNA strands, *Proc. Natl. Acad. Sci. USA,* 89, 1827–1831, 1992.
2. Clark, S. J., Harrison, J., Paul, C. L., and Frommer, M., High sensitivity mapping of methylated cytosines, *Nucl. Acids Res.* 22, 2990–2997, 1994.
3. Riggs, A. D., X inactivation, differentiation and DNA methylation, *Cytogenet. Cell Genet.* 14, 9–25, 1975.
4. Holliday, R. and Pugh, J. E., DNA modification mechanisms and gene activity during development, *Science* 187, 226–232, 1975.
5. Shapiro, R., DiFate, V., and Welcher, M., Deamination of cytosine derivatives by bisulfite. Mechanism of the reaction, *J. Am. Chem. Soc.* 96 (3), 206–212, 1974.
6. Clark, S. J. and Frommer, M., Deamination with NaHSO$_3$ in DNA methylation studies, in *DNA and Nucleoprotein Structure in Vivo*, Saluz, H. P. and Wiebauer, K. R. G., eds., Landes Company, Austin, Texas, 1995, pp. 123–132.
7. Grunau, C., Clark, S. J., and Rosenthal, A., Bisulfite genomic sequencing: systematic investigation of critical experimental parameters, *Nucl. Acids Res.,* 29 (13), E65-5, 2001.
8. Clark, S. J., Millar, D. S., and Molloy, P., Bisulfite methylation analysis of tumor suppressor genes in prostate cancer from fresh and archival tissue samples, *Methods Mol. Med.,* 81, 219–240, 2003.
9. Warnecke, P. M., Stirzaker, C., Song, J., Grunau, C., Melki, J. R., and Clark, S. J., Identification and resolution of artifacts in bisulfite sequencing, *Methods,* 27 (2), 101–107, 2002.
10. Harrison, J., Stirzaker, C., and Clark, S. J., Cytosines adjacent to methylated CpG sites can be partially resistant to conversion in genomic bisulfite sequencing leading to methylation artifacts, *Anal. Biochem.,* 264 (1), 129–132, 1998.
11. Herman, J. G., Graff, J. R., Myohanen, S., Nelkin, B. D., and Baylin, S. B., Methylation-specific PCR: a novel PCR assay for methylation status of CpG islands, *Proc. Natl. Acad. Sci. USA,* 93 (18), 9821–9826, 1996.
12. Paul, C. L. and Clark, S. J., Cytosine methylation: quantitation by automated genomic sequencing and GENESCAN analysis, *BioTechniques,* 21, 126–133, 1996.
13. Warnecke, P. M., Stirzaker, C., Melki, J. R., Millar, D. S., Paul, C. L., and Clark, S. J., Detection and measurement of PCR bias in quantitative methylation analysis of bisulphite-treated DNA, *Nucl. Acids Res.,* 25 (21), 4422–4426, 1997.
14. Rand, K., Qu, W., Ho, T., Clark, S. J., and Molloy, P., Conversion-specific detection of DNA methylation using real-time polymerase chain reaction (ConLight-MSP) to avoid false positives, *Methods,* 27 (2), 114–120, 2002.

15. Gonzalgo, M. L. and Jones, P. A., Rapid quantitation of methylation differences at specific sites using methylation-sensitive single nucleotide primer extension (Ms-SNuPE), *Nucl. Acids Res.,* 25 (12), 2529–2531, 1997.

5 Bisulfite PCR-Based Techniques for the Study of DNA Methylation

James G. Herman

The explosion of studies involving changes in DNA methylation in the last ten years, particularly in the study of cancer, are largely the result of two factors: the increasing awareness of the importance of epigenetic silencing in cancer and the improvements in the techniques used to determine changes in DNA methylation. In this chapter, we will focus on the specific methods utilizing bisulfite-modified DNA and polymerase chain reaction (PCR), discussing the techniques used and information gained, the most appropriate applications for differing molecular questions, and the limitations inherent to each technique.

The increasing awareness that epigenetic changes could play an important role was initially focused on the silencing of bona fide tumor suppressor genes in cancer. These observations were fueled by the discovery of genes that were genetically altered in both familial and sporadic forms of cancer, the classic tumor suppressor genes such as retinoblastoma,[1] von Hippel-Lindau,[2] INK4a,[3] BRCA1, and others that could also be inactivated by promoter region methylation. Studies in the early 1990s were largely accomplished using restriction enzymes to preferentially cleave unmethylated sequences in the promoter regions of these genes, while leaving methylated sequences uncut.[1–3] Southern blot analysis provided the final readout of the methylation status of these CpG islands, but was limited by two main factors: the restriction site's location was fixed in the sequence and such analyses required large amounts of high molecular weight DNA. The increasingly powerful molecular techniques used for genetic changes, including PCR amplification, plasmid cloning, and gene sequencing, were not available for examining epigenetic changes, because 5-methylcytosine content and location were not maintained by DNA polymerases or in place during plasmid cloning. Techniques combining methylation-sensitive restriction enzymes with PCR have been limited by the problem of incomplete enzyme cleavage of the target sequence. What was needed was a method to "fix" the epigenetic pattern in the DNA sequence prior to the application of amplification techniques.

The breakthrough in this area was the recognition that 5-methylcytosine information could be fixed in the DNA by selective deamination of cytosine to uracil with the protection of 5-methylcytosine from the same deamination,[4] initially forming the basis of genomic bisulfite sequencing,[4] but also forming the basis of all other bisulfite-based techniques. This deamination allows the epigenetic modification of cytosine to be effectively converted into genetic information, which can then be subjected to the same molecular techniques that are used for genetic analysis. Following modification, a specific region, or gene of interest, can be amplified using the polymerase chain reaction. The nature of this amplification divides bisulfite PCR-based techniques into two categories: nonselective amplification or methylation-selective amplification. An overall listing of these techniques is provided in Figure 5.1.

Nonselective methods of amplification are all similar in that these techniques seek to amplify all sequences regardless of methylation status following bisulfite modification. This is accomplished by using primers that do not complement (or complement minimally) to regions containing CpG sites that are potential sites of DNA methylation. Further discussion of primer location and design are given below, but the hallmark of these techniques is the intention to not preferentially amplify either methylated or unmethylated sequences. This nonselective amplification then requires a second analysis to determine the actual methylation changes or patterns,

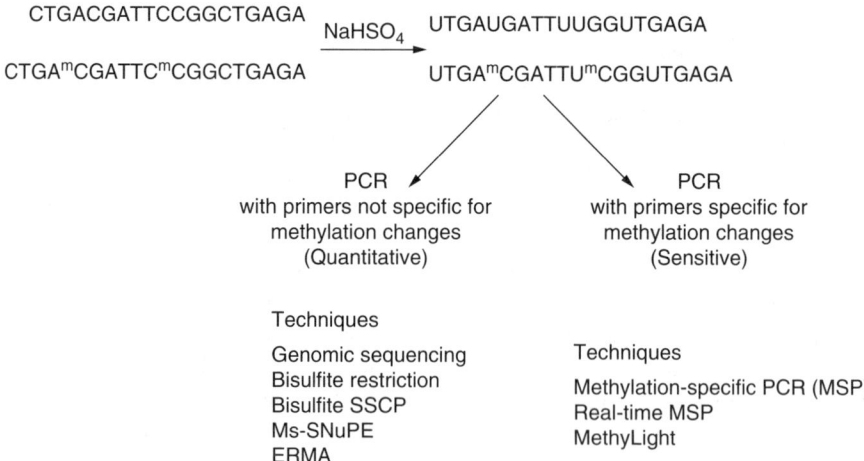

FIGURE 5.1 Methods of determining methylation patterns following bisulfite modification. Shown is a representative sequence that is unmethylated above, and the same sequence with methylation at each CpG site (shown as ^{me}C). Following treatment with sodium bisulfite under denaturing conditions, all cytosines are converted to uracil, but all methylcytosines remain as ^{me}C. Primers are designed for PCR amplification that can avoid these CpG sites (nonbiased approaches) or take advantage of these differences for specific methylation amplification (biased approaches).

Note:
SSCP = single-strand DNA conformation polymorphism.
Ms-SNuPE = methylation-sensitive single nucleotide primer extension.
ERMA = enzymatic regional methylation assay.

and these readouts of the methylation changes are the ways in which these techniques differ from each other.

Genomic bisulfite sequencing, the original bisulfite technique for determining methylation patterns in DNA, remains the most comprehensive method for methylation analysis. Following PCR amplification with nonbiased primers, the methylation status at each CpG site can be determined using either gel-based[4] or, more commonly today, automated sequencing.[5] Two alternatives provide slightly different types of information: direct sequencing from PCR products[5] and the sequencing of individual plasmid clones. The advantage of direct sequencing of a purified PCR product is that the resulting sequence analysis is representative of the sample that was bisulfite treated and amplified; that is, it is a pooled average of the methylation of each CpG site over the amplified region. This approach thus saves the effort, time, and expense of the isolation of multiple plasmid preparations, and the added expense of multiple sequence analyses, and may better represent the overall methylation status of a sample. However, this convenience comes at the cost of a loss of much of the information in methylation patterns contained in the original sequence: all information as to the methylation status of individual copies of DNA is lost in the average methylation result obtained. In contrast, the genomic sequencing of individual plasmid clones provides a detailed analysis of the methylation status of each CpG site on each copy of DNA.[6–8] This allows the evaluation of complex patterns of methylation, a better understanding of the relationship of allelic patterns of methylation, and a better evaluation of mixed patterns of methylation. For example, cloned sequencing can determine whether a segment of DNA with 50% overall methylation has resulted from heterogeneous methylation averaging 50% methylation on all copies of DNA or whether it results from a mixture of complete methylation on some alleles with completely unmethylated DNA on the remaining DNA copies.

However, not all studies require the detailed analysis provided by genomic bisulfite sequencing. In particular, since the silencing of tumor suppressor genes is associated with regional changes in methylation at these CpG islands, an examination of fewer CpG sites within the region may provide representative information about the methylation status of a given CpG island. These techniques, then, allow the examination of more samples, or more genes for each sample with a given amount of resources (time, effort, and expenses). In practice, it is often helpful to combine the use of genomic bisulfite sequencing on some well-studied samples with a simpler analysis on a larger panel of tumors or samples. The simplest approach of these methods is bisulfite restriction analysis.[9] This approach uses the predicted changes in restriction enzyme sites created by the conversion of only unmethylated cytosine to uracil by bisulfite treatment. For example, the sequence CCGA becomes TTGA if unmethylated, but TCGA if the internal (that is, CpG) cytosine is methylated. TCGA is the recognition site for BstUI. This approach uses primers that are identical to those used for bisulfite sequencing, and genomic DNA is bisulfite treated and amplified by the PCR prior to restriction analysis. The restriction products are run on gels to determine the level of methylation, comparing uncut to cut fragments. A more quantitative approach to this restriction analysis is accomplished by the use of radioactive probes.[10] Limitations of this approach are the limited number of sites appropriate for restriction analysis, the inability to chose the specific sites analyzed for methylation analysis,[9] the need for added steps (restriction, gel analysis), and the use of radioactivity for accurate quantitation.[10] However, this approach is relatively simple and can provide some degree of methylation quantitation.

A more precise approach for quantitatively determining methylation status at individual CpG sites has been called Ms-SNuPE (methylation-sensitive single nucleotide primer extension), which is based on bisulfite treatment of DNA followed by single nucleotide primer extension.[11] Genomic DNA is treated with sodium bisulfite, and amplification of the desired target sequence is then performed, using PCR primers specific for bisulfite-converted DNA as for sequencing. The resulting PCR product is then isolated and used as a template for methylation analysis at the CpG sites of interest using primer extension with radioactive C or T. Advantages over alternative methods used for detection of methylation changes include the avoidance of restriction enzymes and the precise quantitation for the methylation status at individual CpG sites.[12] Potential disadvantages include the need for radioactivity and the need for separate amplifications and evaluations for each CpG site examined.

Another accurate method to assess the CpG methylation density of a DNA region in mammalian cells is ERMA (enzymatic regional methylation assay).[13] After bisulfite modification of genomic DNA, the region of interest is PCR amplified with primers similar to sequencing (nonbiased) primers, but modified to contain dam sites (GATC). The purified PCR products are then incubated with 14C-labeled S-adenosyl-L-methionine (SAM) and dam methyltransferase to label the adenine of the dam site as an internal control to standardize DNA quantity. Incubation with 3H-labeled SAM and the bacterial enzyme SssI methyltransferase that methylates all CpG sites for methylation quantification follow this. This method determines an exact measurement of the methylation density of the region studied, essentially averaging the methylation on all copies of DNA amplified by PCR over this region. However, the method again requires radioactivity and does not provide the methylation status of individual CpG sites that MS-SNuPE would provide or the detailed allelic methylation patterns obtained by genomic bisulfite sequencing. It may be best suited for examination of methylation changes induced by demethylating agents.

Finally, other methods analyze the PCR products generated by bisulfite-specific, that is, nonbiased, primers based on conformational changes produced by the sequence differences resulting from bisulfite modification of methylated or unmethylated CpG sites. The combination of bisulfite treatment and PCR-single-strand DNA conformation polymorphism (SSCP) analysis has been proposed for a quantitative methylation assay.[14] In using this technique, PCR products are run under conditions similar to that used to detect mutations in specific genes, in this case, taking advantage of the sequence differences between methylated and unmethylated DNA after bisulfite treatment.[14] Using mixtures of known amounts of methylated and unmethylated DNA, the proportions of methylated DNA can be calculated. More complex patterns observed on SSCP due to heterogeneous patterns of methylation may be more difficult to assess and quantitate because the complex allele differences in methylation may not produce simple banding patterns.[15,16] However, for the lab already familiar with and equipped to perform SSCP analysis, this approach may prove useful. In similar fashion, the products of nonbiased bisulfite PCR can be resolved as differentially methylated sequences by denaturing gradient gel electrophoresis.[17] This method visually displays the degree and heterogeneity of DNA methylation and has demonstrated a high degree of intra- and inter-individual heterogeneity for the *p15* gene in leukemia,[17] which had been found by sequencing analysis.[6,7,18] Denaturing gradient gel electrophoresis (DGGE) and SSCP-based bisulfite techniques yield semi-quantitive methylation information in a given region, but lack either the absolute quantitation of methylation obtained with MS-SNuPE or ERMA or the specific methylation patterns obtained by genomic bisulfite sequencing.

Inherent to the above methods of methylation analysis following bisulfite modification of DNA is the assumption that these sequences are amplified in a nonbiased manner during PCR. Primer design plays an important role in this approach, and the ideal primer for any of the nonbiased PCR approaches incorporates, or overlies, a number of cytosines at non-CpG sites whose modification from cytosine to uracil allows the distinction between bisulfite-unmodified DNA and bisulfite-modified DNA, a point better discussed in Chapter 4. However, it is also important that the primers cover no CpG sites, or at least a minimum number of CpG sites, that would bias toward either unmethylated or methylated sequences (depending on the primer sequence). Alternatively, intentional mismatches at these CpG sites can be incorporated to avoid such bias. If either mixed bases or intentional mismatches are used, these should be located in the 5 portion of the primer to minimize effects on primer annealing and extension. Should either of these constraints prove difficult, it is possible to design PCR primers for either the coding strand or the noncoding strand because bisulfite treatment renders these sequences no longer complementary. For greater detail, one is referred to original descriptions cited above for primer design guidance. However, to achieve any degree of accurate quantitation using any of these methods, one must first be certain that there is not a bias in the initial PCR related to secondary structures that differ between highly methylated and unmethylated sequences following bisulfite treatment, a topic thoroughly examined in a published report.[19]

In contrast to the above methods of methylation analysis following bisulfite treatment, biased approaches utilize the sequence differences produced at CpG sites following bisulfite as the basis for PCR amplification. The earliest and simplest approach to this type of analysis was methylation-specific PCR (MSP).[20] MSP uses the sequence differences resulting from the bisulfite treatment as the basis for primer design. While sequencing and other nonbiased approaches avoid CpG sites in the area to which primers are designed, MSP places primers precisely over these mismatches and utilizes primers specific for the unmethylated and modified sequence and in a separate PCR reaction, the methylated reaction. To achieve maximal discrimination between unmethylated and methylated sequences, differences are mostly placed at the 3 portion of the primer that is most important for primer annealing and extension. The preferential amplification provides two advantages for this approach: (1) the methylation analysis is done when the presence or absence of a PCR product is examined by gel electrophoresis and (2) the specific amplication of methylated sequences allows great sensitivity, and molecular detection approaches using methylation changes are largely based on this type of assay.[21–24] Added sensitivity to simple MSP has been accomplished through the use of fluorescent primers[24] or nested PCR approaches,[23] the latter of which also facilitates the examination of multiple genes in limiting amounts of DNA.[25] The preferential amplification of MSP, however, largely eliminates any useful degree of quantitation, although much like reverse transcription-PCR approaches with limiting cycle number, one might gain gross quantitative information using this approach.

Approaches that restore a degree of quantitation to MSP take advantage of the recent advances in quantitative real-time PCR.[26,27] These approaches monitor the production of PCR product with fluorescent detection, using a variety methods for measurement including simple double-stranded DNA dyes, molecular beacons, or Taqman technology. This type of quantitation must be distinguished from that of the nonbiased approaches above, since what is being quantified here is the amount of DNA having a particular methylation pattern, recognized by the PCR primers or probe, and not the overall level of methylation. Quantitative results of this, and all above quantitative methods, may also be greatly affected by the relative amounts of tumor and normal cells included in the tissue or sample used for DNA extraction.

Finally, a number of critical parameters determine the successful development of these assays and may also markedly influence results and, thus, should be emphasized again. Many follow the phrase used to describe important issues in real estate: "location, location, location." The first location is the site of methylation examined. While one can determine methylation changes for any region of a gene, or for any other region of the genome, methylation-induced gene silencing is associated with methylation changes in the proximal promoter region of a specific gene. Therefore, although each region may differ in CpG content and length, 5 CpG islands should be the starting point for the examination of methylation changes leading to gene silencing. The second location is the overall size of the area PCR amplified following bisulfite treatment. For biased or MSP-based approaches for methylation analysis, short fragments are ideally chosen for amplification, typically 100 to 200 bp. This makes PCR amplifications more robust, allows shorter cycle times, and is needed for Taqman-based approaches due to system constraints. Since the methylation changes examined underlie the primers in any case, there is no need to place the primer far apart. For the nonbiased approaches, in many cases, it would be tempting to amplify large stretches of DNA, since this would allow longer sequencing runs yielding more information in genomic bisulfite sequencing or, for bisulfite restriction analysis, may incorporate more potential restriction sites within the PCR amplicon. However, practical limitations produced by the bisulfite fragmentation of the DNA limit the size of DNA easily amplified during PCR. Realistic PCR products should be less than 500 bp, although with large amounts of good-quality DNA this may be increased somewhat. Finally, the third location is that of the PCR primers. While this separates the biased from nonbiased approaches, it also is critical for assuring PCR amplification, as well as avoiding amplification of bisulfite-blocked DNA, reducing PCR bias in nonbiased approaches, and assuring specific methylation amplification in biased bisulfite approaches.

REFERENCES

1. Ohtani-Fujita, N., Fujita, T., Aoike, A., Osifchin, N. E., Robbins, P. D., and Sakai, T., CpG methylation inactivates the promoter activity of the human retinoblastoma tumor-suppressor gene, *Oncogene,* 8, 1063–1067, 1993.
2. Herman, J. G., Latif, F., Weng, Y., Lerman, M. I., Zbar, B., Liu, S., Samid, D., Duan, D. S., Gnarra, J. R., Linehan, W. M., and Baylin, S. B., Silencing of the VHL tumor-suppressor gene by DNA methylation in renal carcinoma, *Proc. Natl. Acad. Sci. USA,* 91, 9700–9704, 1994.
3. Merlo, A., Herman, J. G., Mao, L., Lee, D. J., Gabrielson, E., Burger, P. C., Baylin, S. B., and Sidransky, D., 5' CPG island methylation is associated with transcriptional silencing of the tumour suppressor P16/CDKN2/MTS1 in human cancers, *Nature Medicine,* 1, 686–692, 1995.
4. Frommer, M., McDonald, L. E., Millar, D. S., Collis, C. M., Watt, F., Grigg, G. W., Molloy, P. L., and Paul, C. L., A genomic sequencing protocol that yields a positive display of 5-methylcytosine residues in individual DNA strands, *Proc. Natl. Acad. Sci. USA,* 89, 1827–1831, 1992.
5. Myohanen, S., Wahlfors, J., and Janne, J., Automated fluorescent genomic sequencing as applied to the methylation analysis of the human ornithine decarboxylase gene, *DNA Seq.,* 5, 1–8, 1994.
6. Dodge, J. E., List, A. F., and Futscher, B. W., Selective variegated methylation of the p15 CpG island in acute myeloid leukemia, *Int. J. Cancer,* 78, 561–567, 1998.
7. Cameron, E. E., Baylin, S. B., and Herman, J. G., p15(INK4B) CpG island methylation in primary acute leukemia is heterogeneous and suggests density as a critical factor for transcriptional silencing, *Blood,* 94 (7), 2445–2451, 1999.
8. Stirzaker, C., Millar, D. S., Paul, C. L., Warnecke, P. M., Harrison, J., Vincent, P. C., Frommer, M., and Clark, S. J., Extensive DNA methylation spanning the Rb promoter in retinoblastoma tumors, *Cancer Res.,* 57, 2229–2237, 1997.
9. Sadri, R. and Hornsby, P. J., Rapid analysis of DNA methylation using new restriction enzyme sites created by bisulfite modification, *Nucleic Acids Res.,* 24, 5058–5059, 1996.
10. Xiong, Z. and Laird, P. W., COBRA: a sensitive and quantitative DNA methylation assay, *Nucleic Acids Res.,* 25, 2532–2534, 1997.
11. Gonzalgo, M. L. and Jones, P. A., Rapid quantitation of methylation differences at specific sites using methylation-sensitive single nucleotide primer extension (Ms-SNuPE), *Nucleic Acids Res.,* 25, 2529–2531, 1997.
12. Daskalakis, M., Nguyen, T. T., Nguyen, C., Guldberg, P., Kohler, G., Wijermans, P., Jones, P. A., and Lubbert, M., Demethylation of a hypermethylated P15/INK4B gene in patients with myelodysplastic syndrome by 5-aza-2-deoxycytidine (decitabine) treatment, *Blood,* 100 (8), 2957–2964, 2002.
13. Galm, O., Rountree, M. R., Bachman, K. E., Jair, K. W., Baylin, S. B., and Herman, J. G., Enzymatic regional methylation assay: a novel method to quantify regional CpG methylation density, *Genome Res.,* 12, 153–157, 2002.
14. Maekawa, M., Sugano, K., Kashiwabara, H., Ushiama, M., Fujita, S., Yoshimori, M., and Kakizoe, T., DNA methylation analysis using bisulfite treatment and PCR-single-strand conformation polymorphism in colorectal cancer showing microsatellite instability, *Biochem. Biophys. Res. Commun.,* 262, 671–676, 1999.

15. Suzuki, H., Itoh, F., Toyota, M., Kikuchi, T., Kakiuchi, H., Hinoda, Y., and Imai, K., Quantitative DNA methylation analysis by fluorescent polymerase chain reaction single-strand conformation polymorphism using an automated DNA sequencer, *Electrophoresis*, 21 (5), 904–908, 2000.

16. Bian, Y. S., Yan, P., Osterheld, M. C., Fontolliet, C., and Benhattar, J., Promoter methylation analysis on microdissected paraffin-embedded tissues using bisulfite treatment and PCR-SSCP, *Biotechniques*, 30 (1), 66–72, 2001.

17. Aggerholm, A., Guldberg, P., Hokland, M., and Hokland, P., Extensive intra- and interindividual heterogeneity of p15INK4B methylation in acute myeloid leukemia, *Cancer Res.*, 59, 436–441, 1999.

18. Melki, J. R., Vincent, P. C., and Clark, S. J., Concurrent DNA hypermethylation of multiple genes in acute myeloid leukemia, *Cancer Res.*, 59, 3730–3740, 1999.

19. Warnecke, P. M., Stirzaker, C., Melki, J. R., Millar, D. S., Paul, C. L., and Clark, S. J., Detection and measurement of PCR bias in quantitative methylation analysis of bisulphite-treated DNA, *Nucleic Acids Res.*, 25, 4422–4426, 1997.

20. Herman, J. G., Graff, J. R., Myohanen, S., Nelkin, B. D., and Baylin, S. B., Methylation-specific PCR: a novel PCR assay for methylation status of CpG islands, *Proc. Natl. Acad. Sci. USA*, 93, 9821–9826, 1996.

21. Belinsky, S. A., Nikula, K. J., Palmisano, W. A., Michels, R., Saccomanno, G., Gabrielson, E., Baylin, S. B., and Herman, J. G., Aberrant methylation of p16(INK4a) is an early event in lung cancer and a potential biomarker for early diagnosis, *Proc. Natl. Acad. Sci. USA*, 95, 11891–11896, 1998.

22. Ahrendt, S. A., Chow, J. T., Xu, L. H., Yang, S. C., Eisenberger, C. F., Esteller, M., Herman, J. G., Wu, L., Decker, P. A., Jen, J., and Sidransky, D., Molecular detection of tumor cells in bronchoalveolar lavage fluid from patients with early stage lung cancer, *J. Natl. Cancer Inst.*, 91, 332–339, 1999.

23. Palmisano, W. A., Divine, K. K., Saccomanno, G., Gilliland, F. D., Baylin, S. B., Herman, J. G., and Belinsky, S. A., Predicting lung cancer by detecting aberrant promoter methylation in sputum, *Cancer Res.*, 60, 5954–5958, 2000.

24. Goessl, C., Muller, M., Heicappell, R., Krause, H., Straub, B., Schrader, M., and Miller, K., DNA-based detection of prostate cancer in urine after prostatic massage, *Urology*, 58, 335–358, 2001.

25. House, M. G., Guo, M., Iacobuzio-Donahue, C., and Herman, J. G., Molecular progression of promoter methylation in intraductal papillary mucinous neoplasms (IPMN) of the pancreas, *Carcinogenesis*, 24 (2), 193–198, 2003.

26. Lo, Y. M., Wong, I. H., Zhang, J., Tein, M. S., Ng, M. H., and Hjelm, N. M., Quantitative analysis of aberrant p16 methylation using real-time quantitative methylation-specific polymerase chain reaction, *Cancer Res.*, 59, 3899–3903, 1999.

27. Eads, C. A., Danenberg, K. D., Kawakami, K., Saltz, L. B., Blake, C., Shibata, D., Danenberg, P. V., and Laird, P. W., MethyLight: a high-throughput assay to measure DNA methylation, *Nucleic Acids Res.*, 28, e32, 2000.

6 Microarray Analysis of DNA Methylation Targets Identified by Methyl-CpG-Binding Proteins

Masahiko Shiraishi, Michael W.-Y. Chan, and Tim H.-M. Huang

CONTENTS

ABSTRACT

We propose the combined use of methyl-CpG-binding domain (MBD) column chromatography and microarray technologies for interrogating cancer genomes. The MBD column chromatograph is based on the affinity binding of methylated CpG island targets to individual MBD proteins and is used to selectively elute bound DNA fragments. The approach is shown to be effective for identifying a large number of methylated loci in the cancer genome. To increase throughput, cloned DNA fragments can further be spotted on microarray slides for differential methylation hybridization. Fluorescently labeled amplicons, which represent different pools of methylated DNA fragments in tumor and control samples, are prepared and co-hybridized onto a microarray slide. Methylated loci are identified based on their differential fluorescence intensities in tumors relative to the control samples. Hierarchical clustering algorithms are further employed to segregate tumor subgroups showing similar methylation profiles. Such a microarray-based analysis, which uses cloned methylated loci by the MBD technology, can routinely be used for clinical diagnosis of specific cancer types.

INTRODUCTION

DNA methylation is a well-recognized epigenetic change that regulates transcriptional silencing in the cancer genome [1]. It takes place by the addition of a methyl group to the fifth carbon position of a cytosine residue in CpG dinucleotides frequently located at the 5-end GC-rich CpG islands of genes [1]. Increasing evidence indicates that in addition to this epigenetic change, other DNA-binding proteins participate in this transcriptional repression, resulting in changes of chromatin structure in and around the promoter regions of a gene target [2]. One family of such proteins is methyl-CpG-binding domain (MBD) proteins [2]. With the exception of MBD4, which is involved in DNA repair, four MBD family members, MeCP2, MBD1, MBD2, and MBD3, are shown to couple to the methylated regions that are transcriptionally inactive in the genome [3].

To further investigate epigenetic changes in cancer, several genome-wide approaches based on enzymatic restriction or bisulfite treatment of methylation targets have been described [4–8]. One other unique technique is MBD column chromatography, which relies on the affinity binding of methylated DNA targets to individual MBD proteins [9–11]. In this approach, the polypeptide corresponding to a methyl-CpG-binding domain is attached to a solid support. The stoichiometry of the column matrix allows for specific binding to methylated CpG sequences, and bound DNA fragments are eluted in appropriate salt fractions. This MBD column chromatography approach can be used effectively to isolate methylated CpG islands in test samples and is advantageous to simultaneously assess methylation profiles of multiple loci in the cancer genome. Here we propose the combined use of this technology with the microarray-based differential methylation hybridization (DMH) [6,12] for high-throughput analysis of cancer-specific DNA methylation. The DNA fragments cloned by the MBD column method can be used to construct a focused DMH microarray for routine assessment of DNA methylation in clinical specimens.

METHYL-CPG-BINDING DOMAIN
COLUMN CHROMATOGRAPHY

RATIONALE AND STRATEGY

At present, MeCP2 is the most commonly used protein for preparing MBD columns [13]. This protein binds specifically to a symmetrically methylated CpG (mCpG) sequence, but not to a hemimethylated or nonmethylated CpG sequence [14]. The MBD column is an affinity matrix that contains a polypeptide corresponding to the methyl-CpG-binding domain of MeCP2, which is attached to a solid support (Figure 6.1) [13]. The stoichiometry of the binding is one polypeptide to one mCpG sequence [15]. Hence, DNA fragments that contain many mCpG sequences have high affinity to the column, while those having the same nucleotide sequence of a fewer number of mCpGs have low affinity to the column. The electrostatic interaction between polypeptide and DNA is disrupted by the addition of salt (e.g., sodium chloride). High concentrations of salt can be used to elute highly methylated DNA fragments that bind tightly to the column, while nonmethylated DNA fragments are eluted in low-salt fractions. Thus, separation of DNA fragments primarily on the basis of the number of mCpGs can be achieved. As such, this approach would be the method of choice when an overall methylation assessment of DNA fragments is intended, but it is limited for the analysis of detailed methylation patterns at the individual CpG dinucleotide level.

EXPERIMENTAL DESIGNS

MBD Column Preparation

The HMBD polypeptide, which is comprised of the methyl-CpG-binding domain of the rat MeCP2 protein (85 amino acids) and a histidine tag (6 consecutive histidine residues) at its N-terminal, has been used to prepare an MBD affinity column [13]. The polypeptide is expressed in bacteria via an expression vector construct [16].

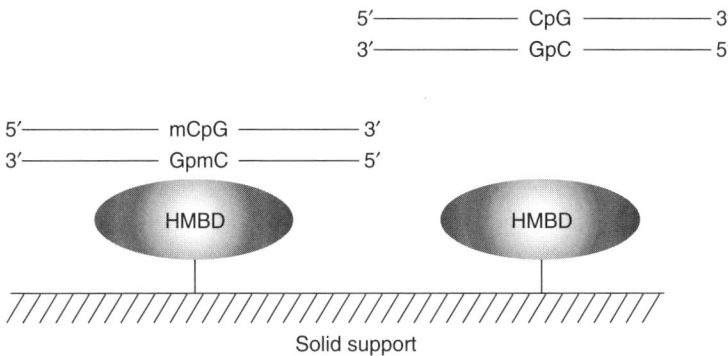

FIGURE 6.1 A schematic illustration of an MBD column. The polypeptide HMBD attached to a solid support binds to a symmetrically methylated CpG sequence at a 1:1 ratio, but not to a nonmethylated CpG sequence.

Because HMBD is basic (predicted pH 9.75), it can be partially purified from bacteria lysate by cation exchange column chromatography. The partially purified product is then passed through an Ni^{2+}-nitrilotriacetic acid agarose column (QIAGEN, 30210; Bio-Rad, Poly-Prep column, 731-1550, 0.8 cm in diameter) to produce an MBD column. The HMBD polypeptide specifically binds to the column through the histidine tag. Typically 30 to 50 mg of polypeptide can be loaded on 1 ml of beads. This amount of polypeptide can be obtained from 1 l of bacterial culture with appropriate induction. In our experience, up to 50 µg of digested human genomic DNA can be loaded and appropriately separated with this size column.

DNA Sample Elution

In order to permit good separation, genomic DNA samples are digested with an appropriate restriction endonuclease. If the target of the analysis is a CpG island, Tsp509I (AATT) or MseI (TTAA) are the enzymes of choice since these sites appear infrequently within a CpG island, and, hence, it is expected that digestion with these enzymes renders the integrity of a CpG island relatively intact. Furthermore, these restriction fragments are amenable to cloning since the recognition sites are compatible with EcoRI and NdeI sites, respectively, which are represented in many cloning vectors.

A linear or stepwise gradient of sodium chloride is applied to elute bound DNA. Since the HMBD polypeptide is basic, nonspecific charge–charge interaction between the polypeptide and DNA can result in the retention of nonmethylated DNA in the column and, as a consequence, their elution at a substantial salt concentration. Detection of target DNAs in the eluate, however, can be achieved either by PCR or Southern hybridization (see examples in Figure 6.2). Since the strength of the interaction is affected by the amount of the bound protein, the salt concentration to elute a specific DNA fragment having a defined number of mCpGs cannot be unambiguously determined. Different concentrations of sodium chloride are required to elute a certain DNA fragment when different batches of MBD columns are used. The retention capacity of a column decreases after extensive use and, therefore, different elution profiles obtained by using the same column cannot always be directly compared.

APPLICATIONS

We have previously used MBD column chromatography to determine the methylation status of the CDH1 CpG island (Figure 6.2). This CpG island locus is not methylated in normal somatic tissue DNA, but is methylated in a gastric cancer cell line MKN1. Tsp509I-digested fragments derived from normal tissue DNA or MKN1 have been analyzed by MBD column chromatography. In DNA from normal somatic tissue, corresponding DNA fragments have been detected in low-salt fractions around fraction 16 (0.52 M NaCl) (Figures 6.2A and 6.2B). In contrast, the CDH1 CpG island fragment has been detected in high-salt fractions 30–36 (0.72–0.80 M NaCl) in the MKN1 DNA sample (Figures 6.2A and 6.2C). These results indicate that nonmethylated and methylated CpG islands can be separated by MBD column

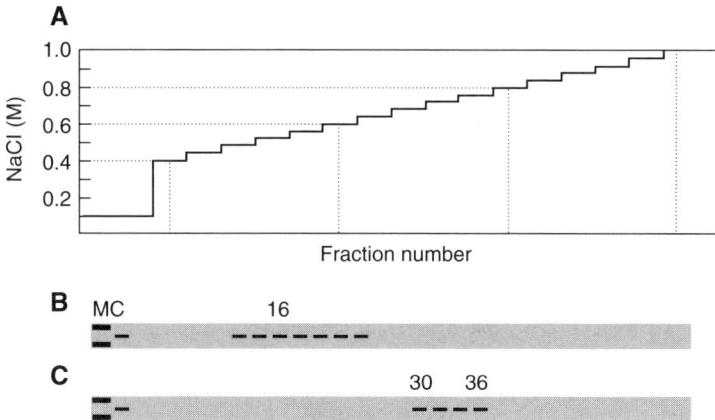

FIGURE 6.2 Separation of genomic DNA fragments by MBD column chromatography. (**A**) A linear gradient of salt (0.4–1.0 M NaCl by 40 mM per step in 20 mM HEPES (pH 7.9, 10% glycerol, 0.1% triton X-100, 0.5 mM phenylmethylsulfonyl fluoride, and 10 mM 2-mercaptoethanol) was applied to elute-bound DNA. 4 ml of elution buffer were applied at each step, and 1.3 ml of aliquot was collected for each fraction. (**B**) 50 μg of *Tsp*509I digests of DNA from human adult male tissue and (**C**) 50 μg of *Tsp*509 I digests of DNA from MKN1 cell line were applied to the column. 300 μl of aliquot were subjected to PCR. M lane: *Hae* III digests of pUC19 DNA; C lane: PCR products from each *Tsp*509I-digested genomic DNA. The details of PCR experiments have been described [17].

chromatography [17,18]. Since the *CDH1* DNA fragment (1.7 kb in length with 126 CpG sites in it) is representative of a CpG island in terms of length and number of CpG sites, it is expected that DNA fragments containing a CpG island will have a similar elution profile. Our results have demonstrated that 80% of *Tsp*509I fragments containing CpG islands give PCR products exclusively in low-salt fractions (Shiraishi, unpublished result).

Occasionally it is desirable to combine eluates into fewer pools to minimize the number of fractions that are analyzed. When an analysis of CpG island methylation is intended, eluates can be pooled in two fractions: a low-salt fraction containing nonmethylated CpG islands and a high-salt fraction containing methylated CpG islands [19,20]. Once properly fractionated, repeated PCR experiments using appropriate PCR primer sets and fractionated DNA as templates permit methylation profiling of many CpG islands [19]. However, as can be easily inferred, PCR products for a DNA fragment containing a nonmethylated CpG island can be also detected in the high-salt fractions when the restriction fragments contain a substantial length of a flanking nonisland region that contains some mCpG sites [20]. An absolute methylation status cannot be determined solely by MBD column chromatography.

MBD column chromatography has also been applied to directly enrich methylated sequences. Many genes are inactivated by CpG island methylation, and it is well accepted that this is an important mechanism of gene silencing in cancer. Some tumor suppressor genes are also inactivated in this manner. In order to perform a comprehensive identification of CpG island methylation in cancer, MBD column chromatography has been applied [10,17]. High-molecular-weight DNA obtained from nine male lung cancer patients has been digested with *Tsp*509I. Highly methylated DNA fragments have been enriched by three rounds of MBD column chromatography and cloned. The cloned fragments have been subject to the segregation of partly melted molecules (SPM) analysis to enrich DNA fragments associated with CpG islands [21]. With the SPM analysis, DNA fragments associated with CpG islands are preferentially retained in a denaturing gradient gel after prolonged electric field exposure. The retained fragments are excised from the gel, cloned, and sequenced. Approximately 6000 plasmid clones have been sequenced, and 700 independent sequences are identified as single-copy candidate CpG island sequence on the basis of the Gardiner-Garden and Frommer's algorithm [22]. Of the 700 unique sequences, 200 CpG island loci are found to be methylated in lung tumors, including those of known tumor-associated genes such as the *HOXA5* gene. These loci are methylation free in normal lung tissue DNA. Thus, this MBD study identifies potential epigenetic biomarkers that can be used for high-throughput methylation analysis by DMH in lung cancer.

POTENTIAL LIMITATIONS AND SOLUTIONS

Although MBD column chromatography is robust, certain DNA fragments show unpredictable elution profiles by this method. For example, DNA fragments with a high density of mCpGs have higher affinity to the column than expected based on the number of mCpGs [9,17]. This phenomenon becomes evident when mCpG densities are greater than 1 per 12 bp in a fragment. These observations suggest that the frequency of the occurrence of mCpG can affect the affinity. The high affinity may be attributed to the protein–protein interaction that potentially strengthens the binding.

A second category of fragments, which have unpredictable elution profiles, are the ones that have sparsely methylated genomic DNA fragments. Some are isolated with DNAs that have high-affinity binding to the column [9,10,23]. When these fragments are cloned and subject to MBD column chromatography after treatment with *Sss*I methylase, they do not show high affinity to the column. Nor do they show high affinity to an MBD column even when they are mixed with digested genomic fragments from which they originate. The reason why such fragments are enriched after repeated MBD column chromatography is unclear.

DIFFERENTIAL METHYLATION HYBRIDIZATION

RATIONALE AND STRATEGY

The methylation status of cloned DNA fragments can be individually verified by PCR-based approaches (e.g., methylation-specific PCR or combined bisulfite restriction analysis) or bisulfite sequencing in primary tumors. However, these conventional approaches may not be practical, especially if a large number of methylated loci are identified by the MBD column chromatography. To increase throughput, we developed DMH for high-throughput analysis of CpG island methylation in cancer. This microarray-based approach has been used to simultaneously profile DNA methylation in breast [6,24,25], ovarian [26], and colorectal [12] tumors, and to identify candidate loci that are associated with clinicopathological parameters of cancer patients. In this regard, the aforementioned 200 loci isolated by MBD column chromatography can be used to construct a DMH microarray subpanel designed solely for the use of lung cancer methylation analysis. The schematic flowchart for DMH is presented in Figure 6.3. PCR products of these loci are arrayed on microscopic slides and cohybridized with fluorescently labeled tumor and control amplicons. One of the key steps in preparing amplicons is the digestion of 4-base methylation-sensitive restriction enzymes (*HpaII* and *BstUI*) in linker-ligated DNA samples. The recognition sites of these restriction enzymes are frequently located within CpG islands. PCR-amplified products are expected to contain different pools of DNA fragments because of the differential methylation status of tumor relative to the control samples. Methylation differences are reflected in Cy5-labeled tumor and Cy3-labeled normal hybridization intensities in the microarray. This can be attributable to a greater abundance of methylated DNA fragments that resist methylation-sensitive digestion in tumor samples and are amplified by linker-PCR, whereas the same unmethylated, allelic fragments are restricted in normal samples and cannot be amplified. Different hierarchical clustering algorithms [27–29] can further be employed to segregate tumor subgroups based on their similar methylation profiles and clinicopathological characteristics.

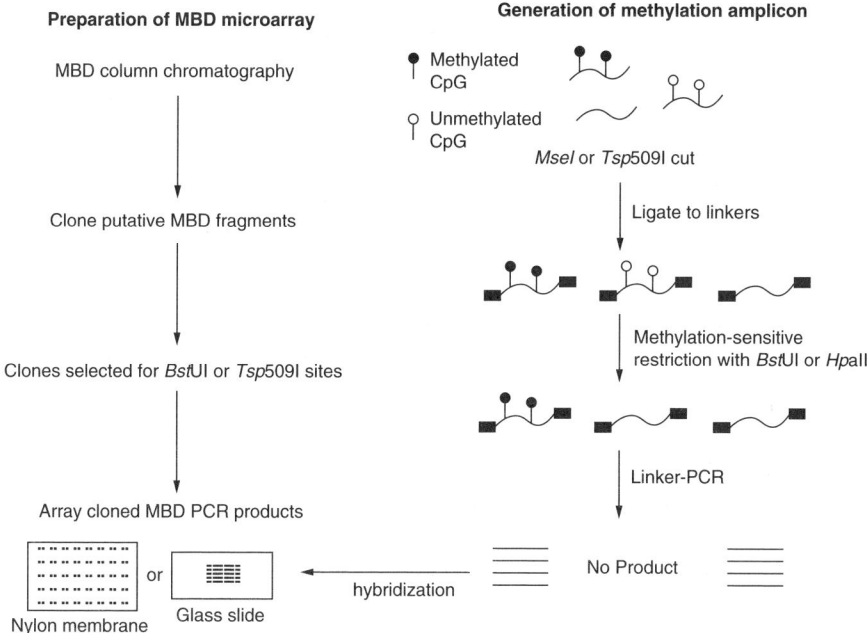

FIGURE 6.3 Schematic flowcharts of differential methylation hybridization. The diagram illustrates the preparation of MBD clones for microarraying and the generation of methylation amplicons, which are used as hybridization probes (see further description in the text).

Experimental Designs

Construction of MBD Microarray

PCR is conducted to individually amplify *Tsp*509I or *Mse*I fragments of the 200 MBD loci with flanking vector primers. The expected sizes of these PCR products are verified by agarose gel electrophoresis. Locus identity can also be verified by DNA sequencing. In addition, 20 *Tsp*509I or *Mse*I fragments from various chromosome locations will be included as controls. These control spots do not contain internal methylation-sensitive sites *Hpa*II or *Bst*UI to be tested by DMH. In the presence of 20% DMSO, purified PCR products (0.1 µg/µl) in triplicate are arrayed onto polylysine-coated microscopic slides by the Affymetrix/GMS 417 microarrayer. The microarrayer can deposit microdots (~0.02 µl per spot and 150 µm in diameter) on a total of 42 slides per run. Prior to microarray hybridization, DNA is denatured and printed slides are crosslinked and postprocessed as described by DeRisi et al. (http://www.microarrays.org).

Preparation of Methylated Amplicon

To prepare methylated amplicons, genomic DNA (1 µg) will be restricted with *Mse*I or *Tsp*509I, the 4-base cutters that restrict bulk DNA into small fragments (<200 bp); its recognition site (TTAA) rarely occurs in GC-rich regions and, thus, most GC-rich CpG islands remain intact after this restriction [13,21]. The restricted CpG island fragments are expected to match the *Mse*I- or *Tsp*509I-digested inserts originally used in the construction of the DMH microarray panel. The cleaved ends of DNA are ligated to unphosphorylated linkers. The ligated DNA is then restricted consecutively with two 4-base methylation-sensitive endonucleases, *Bst*UI and *Hpa*II. Genomic fragments containing methylated sites are protected from the digestion and can be amplified by PCR, whereas genomic fragments containing unmethylated sites are cut, and cannot be amplified in the test sample. Low amplification cycles (20 to 25 cycles) are used for linker-PCR. This approach yields sufficient PCR products for single or low-copy numbers of CpG island loci, but would not overamplify unrestricted repetitive sequences in the ligated DNA. The amplified products are purified, and methylated amplicons (3 to 4 µg each) are fluorescently labeled with Cy5 (tumor) or Cy3 (normal). These labeled amplicons are cohybridized to a DMH microarray slide.

Microarray Data Acquisition

Hybridized microarray slides are scanned and the data generated by the scanning software are exported in a spreadsheet format. The hybridization output is the measured intensities of fluorescence reporters pseudo-colored with red. Positive hybridization signals, which are indicative of CpG island methylation, are scored and exported to an Excel spreadsheet. CpG island loci whose signal intensities are slightly above the background or are devoid of hybridization signals represent the unmethylated loci in the test sample; their genomic fragments are restricted away by the methylation-sensitive endonucleases prior to linker-PCR. Because signal intensities of hybridized targets vary due to different amounts of DNA dotted onto microscopic slides, the signal intensity of each spot will be normalized with that of the internal controls (i.e., 20 MseI or Tsp509I fragments containing no test methylation-sensitive sites). A positive signal for CpG island methylation is scored when a cutoff intensity [(signal – background) × normalized factor] is greater than 20 to 50 (arbitrary units) in triplicated spots of an arrayed probe [30]. It is possible that some positive controls may exhibit greater adjusted hybridization signals than the rest due to an increase in copy numbers of the corresponding loci in the test tumor samples. (Note: hyperdiploidy may be observed in tumor samples.) In this case, not all positive controls can be used for normalization. Based on experience gained from our preliminary studies, the normalization factor will be derived from ~50% of the positive controls. For quality control, it is necessary to periodically conduct a self-hybridization study in which two equal portions of a defined amplicon will be labeled with Cy5 and Cy3, respectively, and cohybridized to a microarray slide. In this case, the adjusted red/green ratio of all the CpG island tags across the microarray panel will be expected to be 1.

POTENTIAL LIMITATIONS AND SOLUTIONS

Normal Stromal and Lymphatic Components in Tumor Samples

This is a common problem for detecting bona fide molecular alterations in clinical specimens. The issue, however, may not be a key concern for the proposed DMH assay. As mentioned earlier, the detection of hypermethylated sequences relies on the presence of amplified products of methylated DNA in tumor amplicons. As such, the presence of a small amount of normal stromal cells (<20%) in macrodissected tumor specimens would not greatly affect the DMH outcome.

Copy-Number Increases vs. True Hypermethylation of the Test Loci in Ovarian Tumors

It is likely that the interrogating sequences do not contain internal methylation-sensitive HpaII or BstUI sites, and the fragments are differentially amplified in tumors relative to normal controls during amplicon preparation. This scenario of copy-number increases in tumors could be detected when we conducted DMH assays using ~8000 CpG island tags randomly selected from the CpG island library CGI [13]. This, however, is not a concern in the MBD application because the internal methylation-sensitive sites of all loci used in the subpanel are individually verified by DNA sequencing.

Incomplete Methylation-Sensitive Restriction during Sample Preparation

Incomplete methylation-sensitive digestion may occur during sample preparation, resulting in false-positive detection of DNA methylation after PCR amplification. Therefore, we have conducted a preliminary test to determine the efficiency of digestion in two CpG island fragments ($BRCA1$ and $p16^{INK4}$) known to be unmethylated in normal blood DNA [30]. Primers flanking a DNA fragment of *BRCA1*, which contains two internal *Hpa*II and three *Bst*UI sites, have been used for PCR amplification. Primers have also been designed for amplification of a DNA fragment in *p16^INK4*. Next, we determined at which cycles of PCR the amplification of undigested DNAs became apparent and found that at >29 cycles of PCR the amplified *BRCA1* or *p16^INK4* fragment can be seen in the digested samples in ethidium bromide-stained gels. These results suggest that the false finding of DNA methylation may occur when high cycles of amplification are used in sample preparation. Consequently, we routinely use fewer PCR cycles (20 to 25 cycles) to prevent unwanted amplification of residual undigested DNA during amplicon preparation.

CONCLUDING REMARKS

Significant efforts have proven the potential utility of genome-based technologies in disease diagnosis. Profiling of transcriptomes and proteomes, which provides vast amounts of molecular information, may eventually replace individual biomarkers currently in use for early detection and prognosis of cancer. This will represent a new paradigm in clinical settings. Therefore, comprehensive analysis of epigenomes represents a new opportunity of biomarker discovery for cancer detection and prognosis. The present study is significant in that novel methylation microarray platforms, which use CpG island loci identified by the MBD technology, can be developed for routine clinical diagnosis of specific tumor types.

Unlike genetic mutations, CpG island methylation is reversible by epigenetic treatments of cancer cells. The combined use of DNA demethylating agents and histone deacetylase inhibitors synergistically reactivates expression of genes silenced by this methylation-mediated mechanism and restores tumor suppressor functions [31,32]. The present microarray-based methylation assay, therefore, will provide useful information regarding which patients will benefit from these reagents as supplements in chemotherapies designed to reverse this aberrant epigenetic condition.

REFERENCES

1. Jones, P. A. and Baylin, S. B., The fundamental role of epigenetic events in cancer, *Nature Rev. Genet.*, 3, 415–428, 2002.
2. Ballestar, E. and Esteller, M., The impact of chromatin in human cancer: linking DNA methylation to gene silencing, *Carcinogenesis*, 23, 1103–1109, 2002.
3. Prokhortchouk, E. and Hendrich, B., Methyl-CpG binding proteins and cancer: are MeCpGs more important than MBDs? *Oncogene*, 21, 5394–5399, 2002.
4. Costello, J. F., Fruhwald, M. C., Smiraglia, D. J., Rush, L. J., Robertson, G. P., Gao, X., Wright, F. A., Feramisco, J. D., Peltomaki, P., Lang, J. C., Schuller, D. E., Yu, L., Bloomfield, C. D., Caligiuri, M. A., Yates, A., Nishikawa, R., Su Huang, H., Petrelli, N. J., Zhang, X., O'Dorisio, M. S., Held, W. A., Cavenee, W. K., and Plass, C., Aberrant CpG-island methylation has non-random and tumour-type-specific patterns, *Nature Genet.*, 24, 132–138, 2000.

5. Ushijima, T., Morimura, K., Hosoya, Y., Okonogi, H., Tatematsu, M., Sugimura, T., and Nagao, M., Establishment of methylation-sensitive-representational difference analysis and isolation of hypo- and hypermethylated genomic fragments in mouse liver tumors, *Proc. Natl. Acad. Sci. USA,* 94, 2284–2289, 1997.

6. Yan, P. S., Chen, C.-M., Shi, H., Rahmatpanah, F., Wei, S. H., Caldwell, C. W., and Huang, T. H.-M., Dissecting complex epigenetic alterations in breast cancer using CpG island microarrays, *Cancer Res.,* 61, 8375–8380, 2001.

7. Toyota, M., Ho, C., Ahuja, N., Jair, K. W., Li, Q., Ohe-Toyota, M., Baylin, S. B., and Issa, J. P., Identification of differentially methylated sequences in colorectal cancer by methylated CpG island amplification, *Cancer Res.,* 59, 2307–2312, 1999.

8. Adorjan, P., Distler, J., Lipscher, E., Model, F., Muller, J., Pelet, C., Braun, A., Florl, A. R., Gutig, D., Grabs, G., Howe, A., Kursar, M., Lesche, R., Leu, E., Lewin, A., Maier, S., Muller, V., Otto, T., Scholz, C., Schulz, W. A., Seifert, H. H., Schwope, I., Ziebarth, H., Berlin, K., Piepenbrock, C., and Olek, A., Tumour class prediction and discovery by microarray-based DNA methylation analysis, *Nucleic Acids Res.,* 30, e21, 2002.

9. Shiraishi, M., Sekiguchi, A., Chuu, Y. H., and Sekiya, T., Tight interaction between densely methylated DNA fragments and the methyl-CpG binding domain of the rat MeCP2 protein attached to a solid support, *Biol. Chem.,* 380, 1127-1131, 1999.

10. Shiraishi, M., Sekiguchi, A., Terry, M. J., Oates, A. J., Miyamoto, Y., Chuu, Y. H., Munakata, M., and Sekiya, T., A comprehensive catalog of CpG islands methylated in human lung adenocarcinomas for the identification of tumor suppressor genes, *Oncogene,* 21, 3804–3813, 2002.

11. Brock, G. J. R., Huang, T. H.-M., Chen, C.-M., and Johnson, K. J., A novel technique for the identification of CpG islands exhibiting altered methylation patterns (ICEAMP), *Nucleic Acids Res.,* 29, e123, 2001.

12. Yan, P. S., Efferth, T., Chen, H. L., Lin, J., Rodel, F., Fuzesi, L., and Huang, T. H.-M., Use of CpG island microarrays to identify colorectal tumors with a high degree of concurrent methylation, *Methods,* 27, 162–169, 2002.

13. Cross, S. H., Charlton, J. A., Nan, X., and Bird, A. P., Purification of CpG islands using a methylated DNA binding column, *Nature Genet.,* 6, 236–244, 1994.

14. Lewis, J. D., Meehan, R. R., Henzel, W. J., Maurer-Fogy, I., Jeppesen, P., Klein, F., and Bird, A., Purification, sequence, and cellular localization of a novel chromosomal protein that binds to methylated DNA, *Cell,* 69, 905–914, 1992.

15. Nan, X., Meehan, R. R., and Bird, A., Dissection of the methyl-CpG binding domain from the chromosomal protein MeCP2, *Nucleic Acids Res.,* 21, 4886–4892, 1993.

16. John, R. M. and Cross, S. H., Gene detection by the identification of CpG islands. In: Birren, B., Green, E. D., Klapholz, S., Myers, R. M., Roskams, J. (Eds.), *Genome analysis volume 2: Detecting genes,* Cold Spring Harbor Laboratory Press, Plainview, 217–285, 1998.

17. Shiraishi, M., Chuu, Y. H., and Sekiya, T., Isolation of DNA fragments associated with methylated CpG islands in human adenocarcinomas of the lung using a methylated DNA binding column and denaturing gradient gel electrophoresis, *Proc. Natl. Acad. Sci. USA,* 96, 2913–2918, 1999.

18. Koizume, S., Tachibana, K., Sekiya, T., Hirohashi, S., and Shiraishi, M., Heterogeneity in the modification and involvement of chromatin components of the CpG island of the silenced human CDH1 gene in cancer cells, *Nucleic Acids Res.,* 30, 4770–4780, 2002.

19. Shiraishi, M., Sekiguchi, A., Oates, A. J., Terry, M. J., and Miyamoto, Y., HOX gene clusters are hotspots of de novo methylation in CpG islands of human lung adeno-carcinomas, *Oncogene,* 21, 3659–3662, 2002.

20. Shiraishi, M., Sekiguchi, A., Oates, A. J., Terry, M. J., Miyamoto, Y., and Sekiya, T., Heterogeneity in the methylation status of genomic DNA fragments demonstrating similar elution profiles in methyl-CpG binding domain column chromatography, *Proc. Jpn. Acad.,* 77 Ser. B, 208–211, 2001.

21. Shiraishi, M., Lerman, L., and Seyika, T., Preferential isolation of DNA fragments associated with CpG islands, *Proc. Natl. Acad. Sci. USA,* 92, 4229–4233, 1995.

22. Gardiner-Garden, M. and Frommer, M., CpG islands in vertebrate genomes, *J. Mol. Biol.,* 196, 261–282, 1987.

23. Selker, E. U., Tountas, N. A., Cross, S. H., Margolin, B. S., Murphy, J. G., Bird, A. P., and Freitag, M.,The methylated component of the Neurospora crassa genome, *Nature,* 422, 893–897, 2003.

24. Huang, T. H.-M., Perry, M. R., and Laux, D. E., Methylation profiling of CpG islands in human breast cancer cells, *Hum. Mol. Genet.,* 8, 459–470, 1999.

25. Yan, P., Perry, M., Laux, D., Asare, A., Caldwell, C., and Huang, T.-M., CpG island arrays: An application toward deciphering epigenetic signatures of breast cancer, *Clin. Cancer Res.,* 6, 1432–1438, 2000.

26. Wei, S. H., Chen, C. M., Strathdee, G., Harnsomburana, J., Shyu, C. R., Rahmatpanah, F., Shi, H., Ng, S. W., Yan, P. S., Nephew, K. P., Brown, R., and Huang, T. H., Methylation microarray analysis of late-stage ovarian carcinomas distinguishes pro-gression-free survival in patients and identifies candidate epigenetic markers, *Clin. Cancer Res.,* 8, 2246–2252, 2002.

27. Eisen, M. B., Spellman, P. T., Brown, P. O., and Botstein, D., Cluster analysis and display of genome-wide expression patterns, *Proc. Natl. Acad. Sci. USA,* 95, 14863–14868, 1998.

28. Sherlock, G., Analysis of large-scale gene expression data, *Curr. Opin. Immunol.,* 12, 201–205, 2000.

29. Tamayo, P., Slonim, D., Mesirov, J., Zhu, Q., Kitareewan, S., Dmitrovsky, E., Lander, E. S., and Golub, T. R., Interpreting patterns of gene expression with self-organizing maps: methods and application to hematopoietic differentiation, *Proc. Natl. Acad. Sci. USA,* 96, 2907–2912, 1999.

30. Chen, C. M., Chen, H. L., Hsiau, T. H.-C., Hsiau, A. H.-A., Shi, H., Brock, G., Wei, S. H., Caldwell, C. W., Yan, P. S., and Huang, T. H.-M., Methylation target array for rapid analysis of CpG island hypermethylation in multiple tissue genomes, *Am. J. Pathol.,* 163, 37–45, 2003.

31. Suzuki, H., Gabrielson, E., Chen, W., Anbazhagan, R., van Engeland, M., Weijenberg, M. P., Herman, J. G., and Baylin, S. B., A genomic screen for genes upregulated by demethylation and histone deacetylase inhibition in human colorectal cancer, *Nature Genet.,* 31, 141–149, 2002.

32. Cameron, E. E., Bachman, K. E., Myohanen, S., Herman, J. G., and Baylin, S. B., Synergy of demethylation and histone deacetylase inhibition in the re-expression of genes silenced in cancer, *Nature Genet.,* 21, 103–107, 1999.

7 Arbitrarily Primed-PCR and Related DNA Methylation Methodologies

Jordi Frigola, Maria F. Paz, Manel Esteller, and Miguel A. Peinado

The dynamics of DNA methylation, and its involvement in multiple biological and pathological processes, requires the use of discovery-based strategies to obtain comprehensive profiles and to reveal specific changes associated with development or disease progression. Due to the extent of genomic methylation, there is no need to identify markers *a priori*, and methodologies based on the screening of large sets of randomly selected anonymous markers may render valuable information at both global and specific levels. The advent of the polymerase chain reaction (PCR) has impacted on the design of multiple molecular biology methodologies and, in the case of DNA methylation, has been widely used not only for analysis of specific sequences, but also in the acquisition of fingerprints representing DNA methylation patterns. PCR-based technologies are frequently chosen because of the requirement of minimal amounts of starting material (usually a few nanograms of genomic DNA) and the relative simplicity of the processes and equipment required. Most PCR-based DNA methylation fingerprinting techniques are adaptations of methodologies previously developed for the analysis of genomes and are based on the sensitivity of restriction endonucleases that are unable to digest methylated CpG sites.

PCR-based DNA fingerprinting methods allow for the generation of large numbers of polymorphic markers while using very small amounts of starting DNA with no need of prior knowledge of the target sequence. These methods are well suited for high-throughput applications used in genomic studies addressing the comparison of two or more samples. Various approaches take advantage of the ubiquitous presence of repetitive elements throughout the genome to produce arbitrary fingerprints. Alu-PCR employs Alu-specific primers [1] and the simple sequence-repeat anchored PCR (SS-PCR) amplification uses primers containing microsatellites and two or three additional nucleotides at the 3 end [2]. Methods based in the use of arbitrary primers, such as arbitrarily primed-PCR (AP-PCR) [3] and random amplified polymorphic DNA (RAPD) [4], are probably the most widely used in studies that range from phylogenetics in prokaryotes and eukaryotes to the genetic analysis of many human diseases. Finally, an alternative method known as amplified fragment length polymorphism (AFLP) produces complex DNA fingerprintings by amplification of restriction fragments ligated to adaptors [5]. In this case, the PCR primers are specific to the adaptor, but include an arbitrary short sequence at the 3 end allowing the selection of a subset of all the fragments.

The various methods reviewed here are summarized in Table 7.1 and share a similar experimental design (Figure 7.1) although, due to differences in the restriction enzyme used, or the selective process during PCR amplification, the outcomes are quite different and are unlikely to reveal the same markers. In all cases, the first step consists of the enzymatic digestion of genomic DNA with a methylation-sensitive restriction endonuclease (Table 7.2). This digestion results in the preservation of DNA fragments containing a methylated restriction site as compared to those that are unmethylated. Often, other restriction endonucleases unrelated to DNA methylation are used in conjunction with the methylation-sensitive enzymes in order to fragment the DNA. The purpose of the additional digestion is to reduce the number of sequences that will amplify independently of their methylation status [6–9].

The second step consists of PCR amplification of sequences meeting specific conditions that are characterisitic of each procedure. Here, two main strategies can be defined: restriction by selective amplification of DNA sequences flanked by sites resulting from the digestion with one of the restriction enzymes [10,11], and restriction by homology of the fragment with the primers used in the AP-PCR [6–8]. Finally, some methods, such as methyl specific-amplified fragment length polymorphism (MS-AFLP) [9] and amplification of interspersed methylated sites (AIMS) [12], use a combination of both strategies. Illustrative schemes of two approaches (MS-AP-PCR and AIMS) are shown in Figures 7.2 and 7.3.

In MS-AP-PCR and MSRF, those sequences containing a methylated site will be preserved (and amplified), as compared to the same sequence containing an unmethylated site that will not be displayed because of the previous digestion with a methylation-sensitive endonuclease [6–8]. A better characterization of the DNA methylation changes is obtained by parallel analysis of the sample without treatment with the endonuclease sensitive to methylation [6–8]. Alternatively, MS-AFLP [9] amplifies the digested (unmethylated) DNA by flanking the fragments with adaptors and using adaptor-specific primers. In this case, the PCR products represent DNA sequences with an unmethylated site. Some methods take advantage of the availability of isoschizomers insensitive to DNA methylation and with different cleavage to produce alternative patterns reflecting methylated and unmethylated targets [10–12].

The sequences that are selected for amplification and its number are highly variable among the different methods. This implies that each strategy is likely to provide markers that are unlikely to be revealed by the others. Additionally, when analyzing complex samples that may contain heterogeneous cell populations, such as those found in a tumor biopsy, the sensitivity limit for *de novo* hypo- or hypermethylations is also variable. For instance, the AIMS method [12] shows sensitivity toward hypermethylation while the MS-AFLP [9] is more efficient in detection of hypomethylation.

TABLE 7.1
Comparative Summary of AP-PCR-Type Methods for the Analysis of DNA Methylation

Method	Endonucleases[1]	Sequence Selection	Displayed Targets	Display Method	Reference
MSRF	MseI + **BstUI**	Sequences flaked by the primer and without an unmethylated BstUI site	Methylated DNA	Gel electrophoresis	(8)
MS-AP-PCR	RsaI + **HpaII/MspI**	Sequences flaked by the primer and without an unmethylated HpaII site	Methylated DNA	Gel electrophoresis	(6)
MS-AP-PCR	MseI + **HpaII/MspI**	Sequences flaked by the primer and without an unmethylated HpaII site	Methylated DNA	Gel electrophoresis	(7)
MS-AFLP	MseI + **NotI**	Sequences flaked by at least one cleaved NotI (unmethylated) site and primer homology	Unmethylated DNA	Gel electrophoresis	(9)
AIMS	**SmaI** + PspAI	Sequences flanked by two cleaved PspA1 (methylated) sites and primer homology	Methylated DNA	Gel electrophoresis	(12)
MS-RDA	HpaII	Sequences falnked by two cleaved HpaII (unmethylated) sites	Unmethylated DNA	RDA	(11)
MCA	**SmaI** + XmaI	Sequences flanked by two cleaved XmaI (methylated) sites	Methylated DNA	RDA	(10)

[1]Methylation-sensitive endonucleases are in bold. Plus sign indicates digestion with multiple enzymes; slash indicates alternative use of the enzyme.

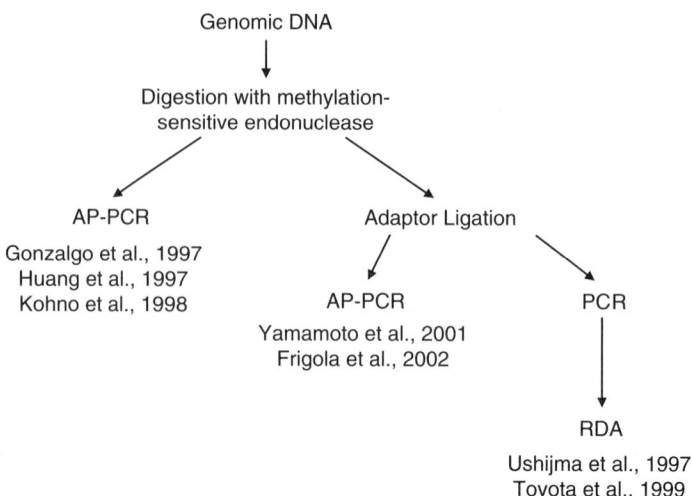

FIGURE 7.1 Scheme of the experimental design shared by the different methods for the detection of differential methylation using AP-PCR-style approaches.

TABLE 7.2
Restriction Endonucleases Used in DNA Methylation Fingerprinting

Sensitive to Methylation			Isoschizomers Insensitive to Methylation	
Enzyme Name	Recognition Sequence	Sites Not Cut[1]	Enzyme Name	Recognition Sequence
BstUI	CG↓CG	ᵐCGᵐCG	–	
HpaII	C↓CGG	CᵐCGG	MspI	C↓CGG
NotI	GC↓GGCCGC	GᵐCGGCᵐCGC	–	
SmaI	CCC↓GGG	CCᵐCGGG	PspAI/XmaI	C↓CCGGG

[1]Only methylated cytosines (ᵐC) at CpG dinucleotides are shown.

Finally, the third step consists of the detection or display of the differences in the methylation profile between two or more samples. Two of the methods perform subtractive representational difference analysis (RDA) [10,11], while in the rest, the generated products are resolved by gel electophoresis, giving rise to DNA fingerprints. The RDA approach is suited to detect relevant markers but is not appropriate for the generation of representative fingerprints. In addition, due to its technical complexity, the comparative analysis of multiple samples is labor intensive and difficult to interpret. By comparison, the fingerprinting methods provide

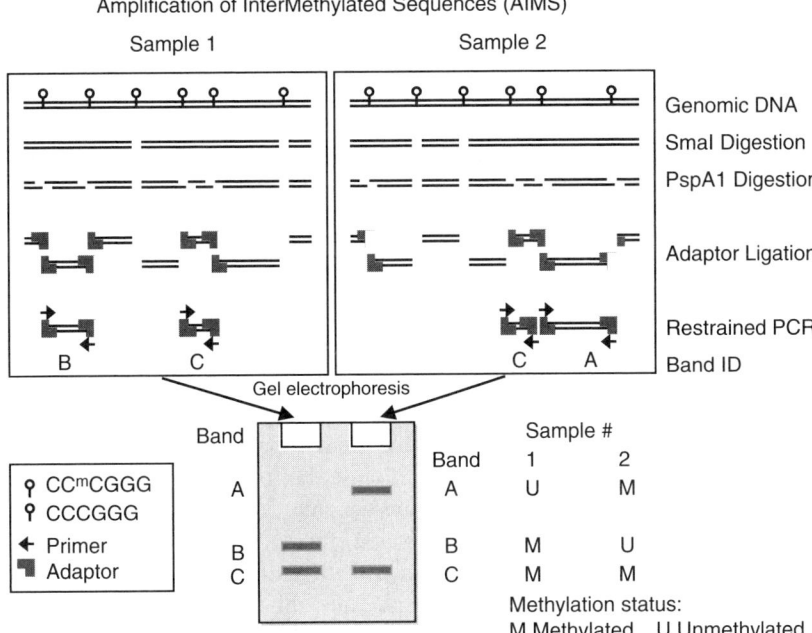

FIGURE 7.2 Diagram of the analysis of differential DNA methylation by methylation-sensitive AP-PCR.

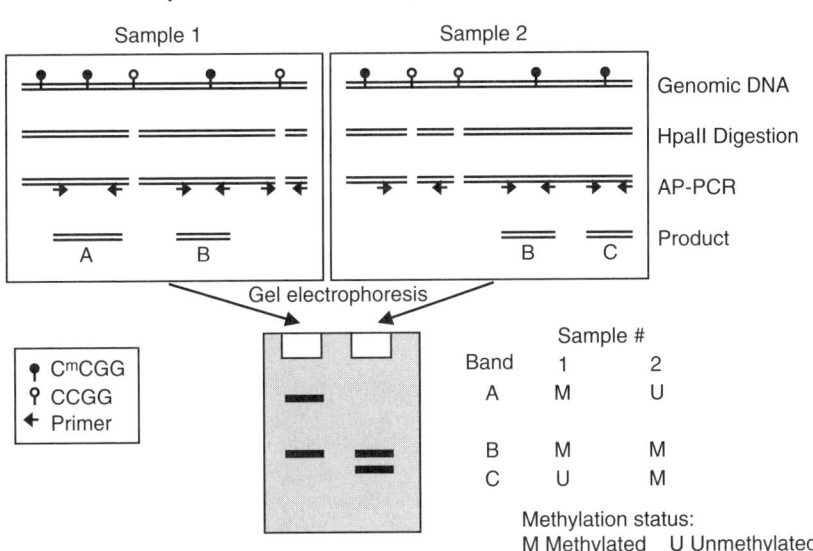

FIGURE 7.3 (**See color insert following page 114**) Diagram of the analysis of differential DNA methylation by amplification of intermethylated sites.

comprehensive profiles that may be used to generate signatures of DNA methylation in addition to the identification of novel markers.

All methods described here have been applied mainly in the field of cancer research, allowing the identification of novel markers of both hypermethylation and hypomethylation in the tumor tissue as compared to the normal cell. A brief summary of some of the applications is shown in Table 7.3.

The last section of this chapter will focus on the AIMS technique [12,14], its applications, and recent improvements. First, in order to enrich the global genomic scanning on GC-rich sequences that undergo methylation changes between different samples, we modified the previously described AIMS approach [12], developing a methyl-enriched AIMS. As in the original method, it allows for the identification of hypo- and hypermethylations by reproducible and readable fingerprints. The isolation, cloning, and sequencing of the fragments that constitute the fingerprints lead to the identification of specific sequences associated with altered methylation patterns. As other nondirected approaches, this method provides precise mapping of methylation aberrations associated with a disease, but is not intended to screen all the genes or all the cytosine residues in the genome. Moreover, the refinement of the technique to address issues of sensitivity and efficiency should provide a valuable tool to generate specific and comprehensive methylation profiles of CpG islands possessing a putative role in carcinogenesis. Thus, this approach has the potential to identify new tumor suppressor genes relevant to the carcinogenic process, since CpG islands are almost always associated with promoters or coding regions of genes [16]. Biotechnology companies, such as Oncomethylome Sciences, are now offering AIMS as a service to find new hypermethylated genes in human cancer.

TABLE 7.3
Applications of DNA Methylation Fingerprinting Methods in Cancer Research[1]

Method	Applications	Reference
MSRF	Detection of hypermethylated CpG islands in human breast cancer	(8)
MS-AP-PCR	Detection of hypermethylated loci in human lung cancer	(7)
MS-AFLP	Genome methylation profiling in human colon cancer	(13)
MS-AFLP	Detection of hypomethylation in satellite DNA in human breast cancer	(9)
AIMS	Detection of hypermethylated/hypomethylated loci in human colorectal cancer	(12)
AIMS	Genome methylation profiling in cancer cells with disrupted DNA methyltransferases	(14)
MS-RDA	Detection of hypermethylated/hypomethylated loci in mouse liver tumors	(11)
MCA-RDA	Detection of methylated CpG islands in human colorectal cancer	(10)
MCA-RDA	Detection of methylated CpG islands in human pancreatic cancer	(15)

[1]This is not a comprehensive list.

The specificity of the technique to analyze the methylation status of the genome relies on the digestion of genomic DNA with methylation-sensitive and methylation-insensitive isoschizomers (*Sma I* or *Cfr9I* and *Xma I*) in parallel for the same sample. Then, a second round of digestions is performed with a single enzyme for all the samples, to obtain homogeneous sticky ends. This double restriction enzyme digestion is used to reduce the number of PCR fragments and potential artefacts. Thus, we are able to join the products of the digestions to an adaptor of known sequence, previously prepared by incubation of two oligonucleotides. The products of ligation are then purified, and thus ready for PCR amplification. This step is performed using different primers or sets of primers [12] consisting of one of the oligonucleotides from the adaptor extended with the overhanging end left by the second enzyme, plus one to four arbitrarily chosen nucleotides. PCR products are diluted in formamide dye buffer, denatured for 3 min at 95°C, loaded into a 6% polyacrylamide 8M urea sequencing gel, and run for 6 h. The gels are dried under vacuum and exposed to an x-ray film. In order to obtain reproducible results, it is interesting to determine previously the minimal amount of adaptor-ligated template DNA, using serial dilutions and performing each reaction twice.

Those fragments showing differences among samples can then be isolated: the appearance of a band denotes hypermethylation, whereas the loss of a band signifies hypomethylation (comparing samples digested with the methylation-sensitive enzyme). Bands with different methylation patterns can be excised from the dried gels, reeluted, and reamplified by PCR under the same conditions. PCR products are then cloned into plasmid vectors and sequenced to ascertain the identity of the isolated band. Sequences are then compared with the genomic databases to identify similarities to any previously known sequence. They can be grouped into any of four categories of DNA sequences: those corresponding to CpG islands (contained or located within), non-CpG islands, unknown or anonymous sequences, and sequences generated from repetitive elements such as SINEs, LINEs, LTRs, Alus, and MIRs [14].

The improvement of the methyl-enriched AIMS technique as a consequence of the double parallel methylation-sensitive digestion described here allows us to overcome potential biases of the AIMS derived from the simple methylation-sensitive digestion. Thus, real methylation alterations can be distinguished from genetic/epigenetic polymorphisms or genomic losses or gains that would otherwise lead to an inaccurate estimation of the CpG islands altered by DNA methylation. This is important, since there is widespread evidence that cancer genomes are genetically and epigenetically unstable.

Therefore, this method can be used for the identification of novel CpG islands that may be progressively methylated during tumorigenesis. Apart from isolating sequences associated with methylation changes in genomic DNA, it is possible to determine quantitatively the relative hypo- and hypermethylation status of genomic DNA. This is given as a ratio between the number of bands with differences of intensity to the total number of comparable bands between the samples (index of methylation).

In order to enable visual detection of methylation changes by AIMS and methyl-enriched AIMS, those anomymous methylation tags can be assigned to different chromosomal regions. It is possible to combine data generated with those approaches with competitive genomic hybridization in order to determine the origin and chromosomal distribution of sequences. Methyl-enriched AIMS products are labeled with SpectrumRed and whole genomic DNA with SpectrumGreen, or vice versa, and then the competitive hybridization of both against a panel of metaphase chromosomes is performed [12,14]. Using this approach, AIMS has identified genes undergoing methylation-mediated silencing in chromosomal regions known to be involved in cancer, such as the loci 19p13 [14].

Finally, it is worth noting the versatility of the AIMS technique. It is a sensitive and accurate approach that can be used to screen for methylation changes in DNA and is suitable for the study of a high number of tags associated with these variations in large series of samples. It has great potential for diverse applications, not only within the field of cancer research, that range from simple comparisons between tumor and normal tissue to combination with other techniques for improved specificity. For example, AIMS can be combined with chromatin immunoprecipitation (ChIP) using immunoprecipitated DNA as the starting material; performing ChIP with antibodies against methyl-CpG-binding proteins and DNA methyltransferases followed by methyl-enriched AIMS will enable the isolation of the genomic targets of these proteins.

REFERENCES

1. Cole, C. G., Goodfellow, P. N., Bobrow, M., and Bentley, D. R. (1991) Generation of novel sequence tagged sites (STSs) from discrete chromosomal regions using Alu-PCR. *Genomics,* 10, 816–826.
2. Zietkiewicz, E., Rafalski, A., and Labuda, D. (1994) Genome fingerprinting by simple sequence repeat (SSR)-anchored polymerase chain reaction amplification. *Genomics,* 20, 176–183.
3. Welsh, J. and McClelland, M. (1990) Fingerprinting genomes using PCR with arbitrary primers. *Nucleic Acids Res.,* 18, 7213–7218.
4. Williams, J. G., Kubelik, A. R., Livak, K. J., Rafalski, J. A., and Tingey, S. V. (1990) DNA polymorphisms amplified by arbitrary primers are useful as genetic markers. *Nucleic Acids Res.,* 18, 6531–6535.
5. Vos, P., Hogers, R., Bleeker, M., Reijans, M., van de Lee, T., Hornes, M., Frijters, A., Pot, J., Peleman, J., Kuiper, M., and Zabeau, M. (1995) AFLP: a new technique for DNA fingerprinting. *Nucleic Acids Res.,* 23, 4407–4414.
6. Gonzalgo, M. L., Liang, G., Spruck, C. H., Zingg, J. M., Rideout, W. M., and Jones, P. A. (1997) Identification and characterization of differentially methylated regions of genomic DNA by methylation-sensitive arbitrarily primed PCR. *Cancer Res.,* 57, 594–599.
7. Kohno, T., Kawanishi, M., Inazawa, J., and Yokota, J. (1998) Identification of CpG islands hypermethylated in human lung cancer by the arbitrarily primed-PCR method. *Hum. Genet.,* 102, 258–264.
8. Huang, T. H., Laux, D. E., Hamlin, B. C., Tran, P., Tran, H., and Lubahn, D. B. (1997) Identification of DNA methylation markers for human breast carcinomas using the methylation-sensitive restriction fingerprinting technique. *Cancer Res.,* 57, 1030–1034.

9. Yamamoto, F., Yamamoto, M., Soto, J. L., Kojima, E., Wang, E. N., Perucho, M., Sekiya, T., and Yamanaka, H. (2001) Notl-Msel methylation-sensitive amplied fragment length polymorhism for DNA methylation analysis of human cancers. *Electrophoresis*, 22, 1946–1956.
10. Toyota, M., Ho, C., Ahuja, N., Jair, K. W., Li, Q., Ohe-Toyota, M., Baylin, S. B., and Issa, J. P. (1999) Identification of differentially methylated sequences in colorectal cancer by methylated CpG island amplification. *Cancer Res.*, 59, 2307–2312.
11. Ushijima, T., Morimura, K., Hosoya, Y., Okonogi, H., Tatematsu, M., Sugimura, T., and Nagao, M. (1997) Establishment of methylation-sensitive-representational difference analysis and isolation of hypo- and hypermethylated genomic fragments in mouse liver tumors. *Proc Natl. Acad. Sci. USA*, 94, 2284–2289.
12. Frigola, J., Ribas, M., Risques, R. A., and Peinado, M. A. (2002) Methylome profiling of cancer cells by amplification of inter- methylated sites (AIMS). *Nucleic Acids Res.*, 30, e28.
13. Yamashita, K., Dai, T., Dai, Y., Yamamoto, F., and Perucho, M. (2003) Genetics supersedes epigenetics in colon cancer phenotype. *Cancer Cell*, 4, 121–131.
14. Paz, M. F., Wei, S., Cigudosa, J. C., Rodriguez-Perales, S., Peinado, M. A., Huang, T. H., and Esteller, M. (2003) Genetic unmasking of epigenetically silenced tumor suppressor genes in colon cancer cells deficient in DNA methyltransferases. *Hum. Mol. Genet.*, 12, 2209–2219.
15. Ueki, T., Toyota, M., Skinner, H., Walter, K. M., Yeo, C. J., Issa, J. P., Hruban, R. H., and Goggins, M. (2001) Identification and characterization of differentially methylated CpG islands in pancreatic carcinoma. *Cancer Res.*, 61, 8540–8546.
16. Esteller, M. (2002) CpG island hypermethylation and tumor suppressor genes: a booming present, a brighter future. *Oncogene*, 21, 5427–5440.

8 Discovering DNA Methylation Differences with Restriction Landmark Genomic Scanning

Christoph Plass, Laura J. Rush,
Dominic J. Smiraglia, and Joseph F. Costello

CONTENTS

INTRODUCTION

Interrogating genomic DNA for abnormal CpG island methylation events and patterns provides important information on disease processes, particularly tumorigenesis. Methylation of CpG islands in promoters is an epigenetic phenomenon associated with transcriptional silencing, and some degree of abnormal methylation has been found in almost every tumor type analyzed thus far.[1] In cancer studies, the methylation status of one or more genes in a tumor is compared to normal tissue. The presence or absence of gene-specific CpG island methylation in tissues and body fluids can serve as a biomarker for early detection of cancer,[2] provide prognostic information,[3] and advance the overall understanding of tumor biology. Methylation studies can be broadly classified into two groups. The first is a candidate gene approach in which the methylation status of a gene of interest, such as a tumor suppressor gene, is investigated in various tumors. The second strategy employs a more global approach, in which many genes, or loci, throughout the genome are assayed to determine widespread changes and to estimate overall methylation levels.[4]

One global technique, restriction landmark genomic scanning (RLGS), is a two-dimensional electrophoretic display of 1500 to 2000 potential methylation events in one assay.[5] Prior knowledge of the genomic sequence of these loci is not necessary; therefore RLGS is well suited for the discovery of methylation targets that were not previously known to be involved in cancer. The methylation readout is a display of CpG islands based on the ability of methylation-sensitive restriction enzymes to digest the DNA under evaluation. RLGS is often used to compare CpG island methylation in tumor DNA with the corresponding normal tissue, but also detects novel imprinted genes,[6] regions of genomic amplifications[7,8] and deletions,[9,10] and hypomethylation events.[11]

THE RLGS METHOD

Biological material for RLGS consists of high molecular weight genomic DNA digested by a methylation-sensitive, rare-cutting restriction endonuclease having a GC-rich recognition sequence (e.g., *Not*I [GCGGCCGC] or *Asc*I [GGCGCGCC]).[5] Overhangs created by the methylation-sensitive enzyme are filled in with radioactive nucleotides, and "landmarks" corresponding to unmethylated CpG islands are present in the final autoradiograph. Conversely, methylated CpG islands are not digested, and no landmark is present because the fill-in reaction for the methylated site does not take place. DNA is then digested with a second enzyme that fragments the DNA into pieces that can be resolved in an agarose tube gel. The second enzyme can vary, but is often a 6 base pair cutter (e.g., *Eco*RV [GATATC]), and it is not methylation sensitive. Following separation in the 0.8% agarose tube gel, the DNA is digested *in situ* with a frequent cutter, such as *Hin*fI (GANTC), to produce fragments of the appropriate size for resolution on the second-dimension non-denaturing 5% polyacrylamide gel. Figure 8.1 illustrates the common *Not*I–*Eco*RV–*Hin*fI enzyme combination. The substitution of *Asc*I for *Not*I, or *Pst*I or *Pvu*II for the other enzymes, results in the display of additional CpG islands, thereby allowing assessment of a larger percentage of the genome than with only one set.[12]

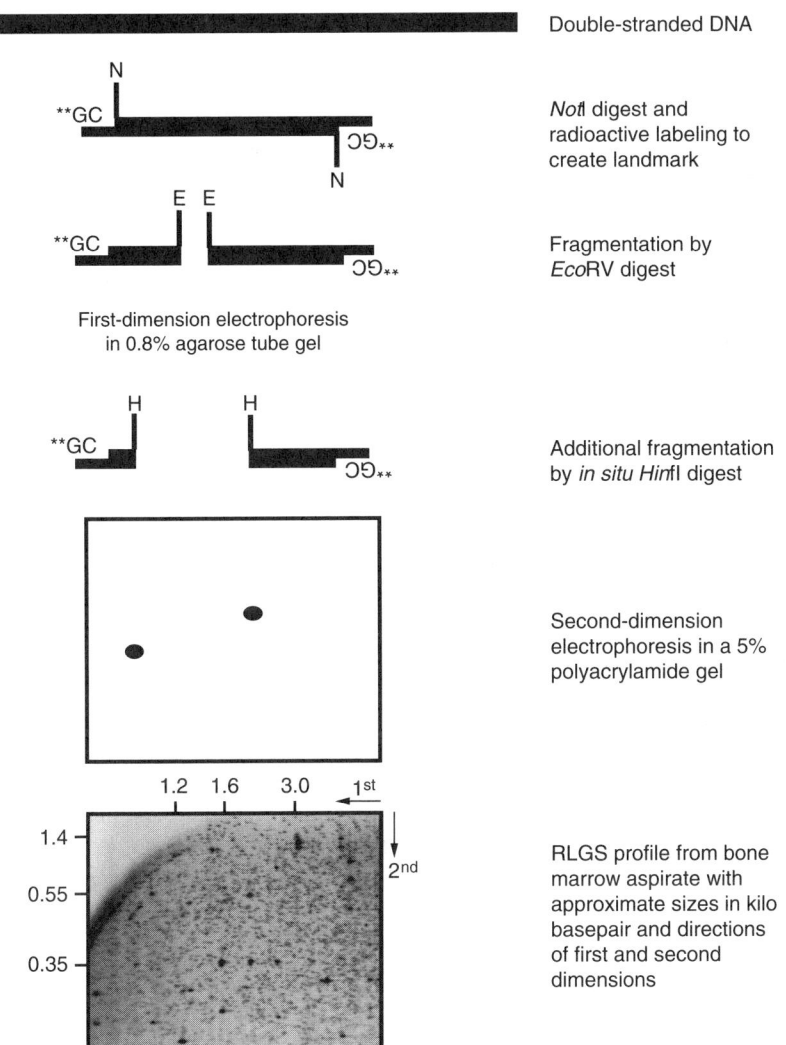

Double-stranded DNA

*Not*I digest and
radioactive labeling to
create landmark

Fragmentation by
*Eco*RV digest

First-dimension electrophoresis
in 0.8% agarose tube gel

Additional fragmentation
by *in situ Hin*fI digest

Second-dimension
electrophoresis in a 5%
polyacrylamide gel

RLGS profile from bone
marrow aspirate with
approximate sizes in kilo
basepair and directions
of first and second
dimensions

FIGURE 8.1 Restriction landmark genomic scanning method. Diagram of the RLGS procedure using the enzymes *Not*I (N), *Eco*RV (E), and *Hin*fI (H). DNA is represented by black rectangles. The *Not*I overhangs are filled in with radioactive nucleotides (**), resulting in a landmark (black oval) on the cartoon representation of a profile. The bottom panel shows an actual RLGS profile from a bone marrow aspirate from a human.

RLGS, like other methylation scanning methods, has inherent strengths and weaknesses. Some features that make RLGS an ideal tool for methylation scanning are

1. The simultaneous assessment of up to 2000 CpG islands in a single assay
2. The high percentage of these CpG islands that are associated with the promoter regions of genes

3. The availability of different enzyme combinations to increase the number of CpG islands assayed
4. The excellent reproducibility between different samples
5. Prior knowledge of genomic sequence not a prerequisite
6. Arrayed plasmid cloning libraries for identification of individual loci
7. High probability of identifying novel methylation targets
8. Ability to quantify methylation alterations and genomic copy number changes
9. No limitations by hybridization kinetics or primer design
10. Applicability to any organism
11. Ability to integrate *in silico* and bioinformatics approaches in CpG island identification
12. Development of automated profile analysis

It must be remembered, however, that RLGS is not an appropriate method for all methylation experiments. For example, 1 microgram of high molecular weight DNA is necessary, which precludes analysis of tissue derived from paraffin blocks or by laser capture microdissection. The procedure is technically challenging and requires equipment not found in most laboratories. Very sensitive PCR-based methods are better suited for detecting low levels of methylation. RLGS provides information about only the methylation status of the "landmark" created by the methylation-sensitive restriction enzyme, not the status of the entire CpG island. Therefore, the usual scenario is that tumor genomes are screened by RLGS to identify methylation targets that are then investigated in more detail with other techniques.

Here we describe how RLGS has been used to discover novel imprinted genes and cancer-related genetic and epigenetic (hyper- and hypomethylation) alterations, and we provide an overview of the latest developments in RLGS profile analysis that capitalize on the human genome draft and state-of-the-art computational biology.

IDENTIFICATION OF SEQUENCES INVOLVED IN THE REGULATION OF GENOMIC IMPRINTING

Most genes are expressed from both the paternal and the maternal alleles; however, a few genes escape this rule. The term "genomic imprinting" describes this phenomenon in which genes are expressed from either the paternally inherited copy or the maternally inherited copy.[13] Fewer than 100 genes out of the total estimated 30,000 genes in the mouse or human genomes are imprinted. Although the reason for the evolution of genomic imprinting is unclear, the importance of imprinted genes is well demonstrated. Deregulation of imprinted gene expression leads to growth impairment, as in the case of Prader-Willi syndrome, Angelman syndrome, or Beckwith-Wiedeman syndrome, and loss of imprinting has been implicated in tumorigenesis.[14–16] Allele-specific DNA methylation in, or close to, sequences that regulate imprinted expression has been found.[17] This characteristic feature seen in all imprinted regions has been used in the past for the identification of novel

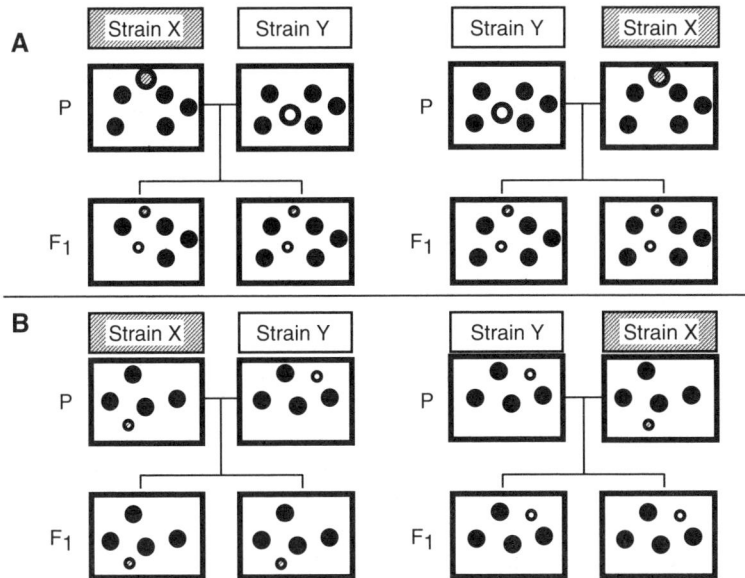

FIGURE 8.2 Identification of imprinted loci by RLGS: (**A**) Transmission of nonimprinted polymorphic RLGS fragments. Polymorphic fragments are present only in either strain X or strain Y profiles, while these polymorphic fragments show haploid intensity in the F_1 animals of both reciprocal crosses between strains X and Y. RLGS patterns of F_1 animals from reciprocal crosses are identical. (**B**) Transmission of imprinted polymorphic sequences in RLGS profiles. RLGS fragments in imprinted regions show either a paternal or maternal specific transmission. Profiles of F_1 animals of reciprocal crosses differ in these fragments.

imprinted genes, or imprinted regions, using the RLGS method. RLGS profiles of inbred mouse strains show varying numbers of strain-specific polymorphic fragments, a prerequisite for the identification of imprinted loci by RLGS.

Such mouse strain-specific differences are seen on an RLGS profile from C57BL/6, which shares about 87% of RLGS fragments with the DBA/2 strain, indicating that each profile has approximately 13% strain-specific RLGS fragments. RLGS profiles of F_1 progeny from a genetic cross of C57BL/6 with DBA/2 animals display all RLGS fragments that are seen in the individual profiles. Fragments found in both strains will be displayed with diploid signal intensity, whereas strain-specific RLGS fragments exhibit haploid or half intensity (Figure 8.2A).[18] While the majority of polymorphic fragments behave the same in a reciprocal cross, RLGS fragments located in imprinted regions do not.[19] First, all restriction landmark sites in imprinted regions would show up as haploid fragments in each of the parental RLGS profiles. Second, a landmark site in an imprinted region would be present only in F_1 animals of one direction, but not in the reciprocal cross (Figure 8.2B). The different transmission patterns between nonimprinted and imprinted polymorphic RLGS fragments have been used in the past for the identification of novel imprinted genes.[20] An RLGS screen of mouse reciprocal crosses led to the identification of eight RLGS candidate loci, termed Irlgs 1–8, that followed a paternal- or maternal-specific transmission.[19]

Molecular analysis of Irlgs2 resulted in the discovery of *U2afbprs*, a gene with homology to the human U2 auxilary factor 35-kDa small subunit on mouse chromosome 11.[21] *U2afbprs* is expressed from the paternal allele only. Interestingly, the human *U2AFBPRS* gene is not imprinted,[22] a feature seen only in a small subset of imprinted genes. Molecular characterization of Irlgs3 resulted in the identification of *Rasgrf1*, a guanine nucleotide exchange factor expressed from the paternal allele in the neonatal brain.[6] Imprinting of the human counterpart of *Rasgrf1* has not yet been described. A cell cycle control gene, ZAC/PLGL1, was identified as an imprinted gene in a screen for novel imprinted genes in humans by comparing parthenogenetic DNAs with those DNAs derived from hydatidiform moles.[23]

ANALYSIS OF CANCER IN HUMANS AND MICE

The reproducible and quantitative nature of RLGS make it well suited for tumor and normal tissue comparisons.[5] Ideally, comparisons are made between patient-matched normal and tumor samples to eliminate the potential effect of genetic polymorphisms on RLGS spot intensity. Tissue-matched or cell-type-matched controls are also useful to distinguish methylation that varies in different normal cells from truly aberrant changes in cancer. The parallel analyses of more than 1000 CpG sites afforded by RLGS is critical for pattern recognition within, and among, cancer types, and for estimating the overall influence of CpG island methylation on the cancer cell genome.[24-26] Human cancer types that have been analyzed by RLGS include leukemias[26] and solid tumors of the brain,[24,27-30] breast,[24] colon,[24] head and neck (including matched primary and metastatic pairs),[25] liver,[31-36] lung,[37,38] meninges,[39] oral cavity,[40] adrenal gland,[41,42] and testis.[43] Comparison of tumor and normal control RLGS profiles has been used to detect alterations in methylation such as aberrant CpG island methylation and loss of methylation in high copy number sequences,[24-27,37,41,44,45] and also copy number changes including deletion,[10,26,28,46] chromosomal gains, and regional amplification.[8,10,28,30,38,47-51] (See Figure 8.3.)

ABERRANT METHYLATION OF CPG ISLANDS

CpG islands on autosomal chromosomes are usually unmethylated, but may become aberrantly methylated in tumors, thus contributing to transcriptional inactivity of the associated gene. Using the candidate gene approach of studying methylation in genes first identified through classic genomic screens, aberrant CpG island methylation has been shown to be associated with silencing of a wide variety of tumor suppressor and cancer-related genes.[1] In contrast, RLGS and other methods have been used to identify novel cancer genes that are primarily affected by aberrant methylation, and are missed by copy number and mutational analysis of tumor DNA.[10,24-26,28,37] On RLGS profiles, aberrant methylation of a CpG site results in a decrease of intensity of the corresponding spot. In many cases, a single copy spot is completely absent from a tumor profile indicating that the methylation (and/or deletion) event has occurred in both alleles of the gene and in the majority of cells in the tumor. This fact suggests that many aberrant methylation events occur early in the tumorigenesis, though this appears to be CpG-island- and tumor-type dependent.

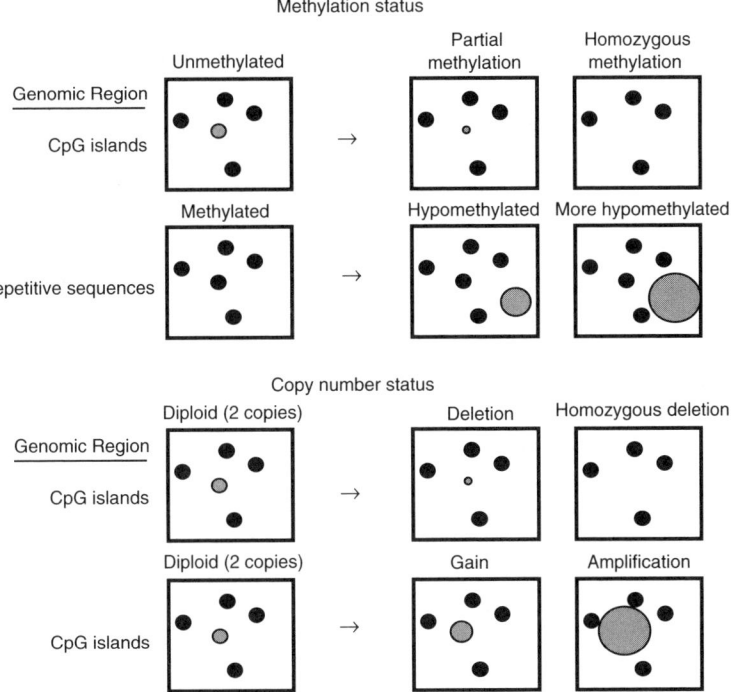

FIGURE 8.3 Methylation and copy number changes detectable by RLGS. Most CpG islands are unmethylated in normal cells (left column of boxes) and, therefore, generate a spot of diploid intensity (gray and black spots in top, left panel). In tumors, CpG islands may become aberrantly methylated (top, right two boxes), resulting in a decrease in spot intensity that is equivalent to the level of methylation. Note that while one spot may decrease in intensity (gray spot), others remain unmethylated (black spots). In contrast, high copy number sequences may be methylated in normal cells, but become hypomethylated to varying degrees in tumors (see gray spot in second row, right two boxes). Copy number changes such as deletion and amplification are also detected by RLGS (lower 2 rows of boxes). The left column depicts RLGS profile close-ups of normal tissue, while the right two columns depict tumor profiles. "Gain" consists of a diploid locus becoming 3 to 4 copies, while loci are considered amplified if the copy number is 5 or greater (gray spot).

The RLGS approach illuminates patterns of methylation that might yield clues to the underlying mechanism of aberrant methylation and tumorigenesis. First, the observation that some CpG islands are preferentially methylated suggests that clonal selection and/or different susceptibilities of CpG islands to hypermethylation may shape the patterns in tumors.[24] An argument in favor of differential susceptibility to aberrant methylation was recently reported.[52] In this study, RLGS analysis of cells overexpressing the DNA methyltransferase DNMT1 was used to identify CpG islands that were frequently methylated or never methylated. Using DNA pattern recognition algorithms and supervised learning techniques on this data, a classification function was derived that is capable of discriminating methylation-prone from methylation-resistant CpG islands. This suggests that there is a component of "susceptibility to methylation" that is DNA-sequence-dependent.[52] Second, the patterns of aberrant methylation across CpG islands displayed on RLGS profiles are clearly nonrandom and tumor-type-specific, suggesting that methylation of specific gene subsets may contribute to the generation of particular cancer types. In addition to the nonrandom nature of aberrant methylation across the genome, acute myeloid leukemias exhibit an overrepresentation of aberrant methylation occurring on chromosome 11.[26] Third, the total number of aberrantly methylated CpG islands in sporadic human tumors was estimated to be in the hundreds from RLGS profiles.[24] These estimates should not be interpreted to reflect an equivalent number of methylation-silenced genes, as some aberrant methylation may not influence gene expression due to the genomic location or methylation density. Fourth, the total number of methylated sites is variable between and, in some cases, within different tumor types, suggesting there may be methylation subtypes within tumors having similar histology.

Aberrant methylation of individual CpG islands discovered by RLGS has been shown to be functionally and/or clinically relevant to tumorigenesis. For example, two aberrantly methylated genes, *SOCS1* in liver tumors[53] and *SLC5A8* in colon cancer,[54] have been shown to possess tumor suppressor activity. Dai et al. identified frequent promoter methylation of bone morphogenetic protein 3B (BMP3B) in non-small-cell lung cancer and demonstrated that BMP3B has growth-suppressive properties in lung cancer cells.[55] In acute myeloid leukemia (AML) patients, methylation of WIT1 is significantly more frequent in refractory AML compared to chemosensitive AML.[56] These and additional studies[35,36] suggest that RLGS is a useful tool for discovering new tumor suppressor genes and may be a rich source of biomarkers for predicting clinical behavior of tumors.

Relative to primary tumors, human tumor cell lines exhibit significantly higher levels of CpG island methylation.[57] RLGS analysis revealed a 5-fold to 93-fold increase in the frequency of CpG island methylation in cell lines compared to their primary tumor counterparts. More than 57% of CpG islands that were methylated in these tumor cell lines were never methylated in primary tumors. This study suggests that a significant proportion of the methylation in cell lines is an intrinsic property of cell lines rather than their malignant tissue of origin.[57]

HYPOMETHYLATION

Loss of methylation in tumors has also been detected using RLGS. Such hypomethylation has been reported for several high copy number sequences that are normally methylated, including pericentromeric sequences and a tandem repeat sequence on chromosome 8q21.[41,45] The demethylation appears as a single new spot with an intensity equivalent to the number of repeats that become demethylated and can exceed 100-fold in some cases. Hypomethylation has been linked to chromosome instability and may be a contributing factor to specific chromosomal abnormalities seen in human cancer. The presence of hypomethylated repeat sequences in hepatocellular carcinoma is significantly associated with tumor recurrence, even in patients with a conventional good prognosis.[36]

REGIONAL DNA AMPLIFICATION

In addition to hypomethylation of high copy number sequences, amplification of chromosomal regions in tumors also results in an RLGS spot of high intensity.[47] In most instances of amplification detected on RLGS profiles, the amplification has been found to increase the intensity of a normally single copy spot, rather than creating a new spot as from hypomethylation. Novel amplified regions (and corresponding candidate genes) identified by RLGS include 7q21 (CDK6) and 1q32.1 (GAC1) in brain tumors, 19q13.1 (e.g., AKT2) in pancreatic cancers, 8p22-23 (Cathepsin B) in esophageal adenocarcinoma, 2p24 (NAG along with N-myc) in neuroblastoma and medulloblastoma, 11q22 (cIAP1 and cIAP2) in lung cancer, and 1p34-35 (L-myc) in ovarian carcinomas.[8,30,38,48–51,58]

INTEGRATED GENETIC AND EPIGENETIC ANALYSIS OF TUMORS

Due to the similar appearance of hypomethylation and gene amplification, as well as CpG island methylation and deletion on RLGS profiles, it is not always possible to distinguish genetic from epigenetic events. To address the relative contribution of copy number changes and methylation alterations to RLGS spot changes, and to tumorigenesis, RLGS analysis was paired with a high-resolution copy number analysis (array comparative genome hybridization) on the same tumor set.[10,28] These studies showed that subsets of gene-associated CpG islands are preferentially affected by convergent methylation and deletion, including genes that exhibit tumor suppressor activity such as *SOCS1,* as well as novel genes such as *COE3* and *ZNF342* that have been missed by traditional nonintegrated approaches. These results also show that most aberrant methylation events are focal and independent of deletions, and aberrant CpG island methylation appears to be far more prevalent than deletion as a source of RLGS spot loss.[10,28]

MOUSE MODELS OF HUMAN CANCER

RLGS analysis of mouse models of human cancer has also identified targets of amplification and aberrant CpG island methylation. For example, RLGS analysis of mouse liver tumors has identified aberrant methylation of *α4 integrin* and *CDKN2A* as well as amplification and overexpression of *cyclin A2*.[46] In addition, other novel targets of aberrant methylation and gene silencing have been identified in mouse tumors, such as *mlt 1* and *Igfbp7*.[59,60] Nonrandom deletions on chromosomes 1, 5, 7, and 13 have also been identified in mouse hepatoma.[9] Thus, RLGS analysis of these tumors indicates that the models recapitulate, in part, the genetic and epigenetic alterations seen in human tumors. These studies also highlight the ease with which RLGS can be applied to any species.

RLGS is, thus, a useful tool for identifying novel cancer genes, understanding the relative contribution of genetic and epigenetic mechanisms in tumorigenesis, and providing a rich source of biomarkers for prediction and prognosis in the cancer clinic. Further advancements in identification of RLGS spots and expansion of the number of sites analyzed will provide considerable impetus for the future application of RLGS to cancer genomes.

RLGS SPOT CLONING

While DNA methylation changes can be detected using RLGS images by virtue of spot loss, the DNA sequence and chromosome location of the DNA fragments represented by RLGS spots are unknown. One way to obtain this information is to clone the RLGS fragments. Cloning RLGS fragments, or simply identifying the sequence, is a critical step that allows us to go from counting methylation events to identifying the targets of hypermethylation. The amount of DNA present in a diploid spot of an RLGS profile is estimated in the 10^{-15}M range. Elution and direct cloning of such low amounts of DNA proved to be very difficult and, therefore, early RLGS fragment cloning was limited to the high copy number RLGS fragments.[30,42] Technical improvements, such as use of the restriction trapper and PCR-mediated cloning methods,[61–63] have allowed for the cloning of some diploid RLGS fragments; however, these methods remain inefficient and have inherent technical problems.

TARGETED LIBRARY CLONING STRATEGY

The cloning of RLGS fragments from human profiles was greatly improved by the creation of an arrayed *Not*I–*Eco*RV boundary library.[64,65] Total genomic DNA was completely digested with *Not*I and *Eco*RV, cloned into a plasmid vector and arrayed in 384-well microtiter plates, each well representing a single clone. Clones from this boundary library were pooled from each of 32 microtiter plates. The pooled clones from each individual plate were labeled and mixed with labeled genomic DNA at the appropriate ratio, such that fragments represented by clones from that pool show an approximate 10-fold enhancement in intensity when the mixture of clone and genomic DNA is run in an RLGS profile. Thirty-two plate pool mixing gels have been prepared in this way. In addition, each of 16 rows (A–P) from all 32 plates were pooled and used to create individual row pool mixing gels. The same has been done for each of 24 columns. Thus, a spot cloning resource of 32 plate mixing gels, 16 row mixing gels, and 24 column mixing gels has been created. The clones from these 32 plates cover approximately 70% of the fragments in a normal RLGS profile from peripheral blood lymphocyte (PBL) DNA.

In order to identify the correct library clone address for one of these fragments, the correct plate, row, and, column mixing gels in which the fragment of interest is enhanced must be determined. Owing to the redundancy of the library, many fragments are represented more than once in this set of 32 plates. This results in finding the fragment of interest enhanced in more than one plate, row, and column mixing gel, creating many different possible combinations, which must be tested to identify the proper address in the library. To identify the correct address, all possible row and column combinations where the spot of interest is enhanced are used in combination with one of the identified plates. The correct clone is identified by determining which clone has an insert size closest to the spot size as predicted by its gel migration distance. To confirm that the correct clone has been identified, plasmid DNA from the clone is used to create a mixing gel to demonstrate that this single clone enhances the correct spot. Numerous reports using this method have been published[7,24,26,30,37,43,44,56–58,64–67] and analysis of these clones provided unequivocal evidence that RLGS preferentially scans CpG islands, and when it can be determined, that these tend to be in the 5' ends of genes.[12,43] Similar cloning libraries have been created for the mouse and rat genomes as well as a second enzyme combination (*Asc*I–*Eco*RV) in human.

NONTARGETED LIBRARY CLONING STRATEGY

In order to clone large numbers of RLGS spots using the boundary libraries previously described, a strategy was devised using mixing gels that would result in the identification of every RLGS spot in the set of gels. To accomplish this it was necessary to eliminate the redundancy in the library by sequence analysis of a portion of the library, followed by rearraying the nonredundant clones. Large-scale sequencing of 10 plates in the library totaling 3840 clones was performed and the sequences were compared to one another. A set of nonredundant clones was identified and rearrayed into 96-well plates. These 96-well nonredundant plates were used to create plate, row, and column mixing gels as described. However, in this set of gels, each spot could be present only once and, therefore, would have only a single plate, row, and column address. Each mixing gel was then carefully analyzed to identify each enhanced spot. This information resulted in the identification of unique library addresses for many RLGS spots.

VIRTUAL RLGS: SPOT SEQUENCE IDENTIFICATION

A virtual RLGS (vRLGS) image is a computer-generated RLGS image constructed from a genome database. Each spot in the virtual image comes from a clearly identified DNA sequence or set of sequences in the human genome database. By matching the spots in the virtual image to spots on a laboratory gel, we can identify the DNA sequence and chromosome location of a given RLGS fragment.

For a *Not*I–*Eco*RV–*Hin*fI RLGS profile, the migration of each RLGS fragment in the first dimension is determined by the size of the *Not*I–*Eco*RV fragments (and the *Not*I–*Not*I fragments without an internal *Eco*RV site). In the second dimension, these fragments are further digested in gel with *Hin*fI and thus the migration in the second dimension is determined by the *Not*I–*Hin*fI fragment sizes. Note that only the *Not*I half-site is labeled, and only the *Not*I–*Hin*fI fragments are ultimately visualized. The size-dependent migration of these fragments in both dimensions can be predicted based upon empirical data. Thus, it should be possible to use bioinformatics to predict what an RLGS profile should look like based upon the publicly available human genome sequence. Such an approach could be used as a new RLGS spot sequence identification strategy. Rouillard et al. recently demonstrated a first generation example of such an approach.[68] A similar method using *in silico* digests of human genome sequence along with chromosome-assignment data[69] was used to identify and confirm the nucleotide sequence of 200 RLGS spots.[10,28]

The location of an RLGS spot depends on the distance the corresponding DNA fragment migrates in the gel, both in first and second dimensions. This migration distance correlates highly to the length of the DNA fragment. Using previously described library RLGS spot cloning techniques, DNA fragment information was identified for 300 spots on a *Not*I–*Eco*RV–*Hin*fI master gel. Spot locations came from a master RLGS profile, used as a standard for all gels.[24] For the *x*-coordinate, a simple linear regression model $D\sim\log(L)$, where D is the distance the DNA fragment migrates and L is the fragment length for both *x* and *y* gel coordinates, was fitted to the data. For the *y*-coordinate, it was observed that the migration pattern of 278 smaller DNA fragments (length <800 bp) is slightly different from that of 22 longer fragments (length 800). Hence, for the *y*-coordinate, a piecewise regression model to derive the prediction formula was developed, which was more accurate than a single regression model. It became clear that the *y*-dimension prediction was not as accurate as the *x*-dimension prediction. To improve prediction of the *y*-coordinate, other feature variables that may contribute higher accuracy to the fitted model were considered. The other feature variables considered were DNA curvature, GC%, and CpG%. Therefore, using this version of vRLGS that incorporates DNA curvature and GC content into the migration models, it is possible to correctly identify the sequences of previously unidentified RLGS spots.

In 1996, Yoshikawa et al. used flow-sorted chromosomes to run chromosome-specific RLGS profiles.[69] These individual chromosome RLGS profiles were put together with each spot labeled to indicate which chromosome it came from. We first compared the chromosome-assigned RLGS profile[69] with our master profile to identify what chromosome each spot of interest was expected to be from. For each of four spots, we used the vRLGS profile to identify the sequence predicted to migrate closest to the actual spot that was from the correct chromosome. These were our set of candidate sequences. We designed PCR primers from each of these candidate sequences and eluted the DNA from a human *Not*I–*Eco*RV–*Hin*fI RLGS profile for all four spots. We then used the primers for the candidate sequence to amplify DNA eluted from the spot in question, as well as a nearby spot that served as a negative control. Accurate prediction of a spot sequence was determined if a strong PCR product was obtained when that spot was used as template, but no product or a very weak product was detected using a nearby spot as template.

Figure 8.4 shows an example of the vRLGS spot sequence verification experiment. The left panel shows the actual RLGS gel at section 3D, while the right panel shows the virtual profile from the same section. The lower panel is a composite of those two images. We predicted the sequence of spots 36, 56, 62, and 68 from section 3D of the master profile based on the virtual spots indicated. In all four cases, the PCR experiment described showed a strong band when the primers from the predicted sequence matched the template DNA used, but no band when DNA from a nearby spot was used as template.

As is clear from Figure 8.4, the virtual pattern does not perfectly match the actual pattern. Spots 56, 62, and 68 are very close, but spot 36 has approximately a spot's width gap between the virtual and actual spots. Also, there are extra spots in the virtual profile. These extra spots most likely arise from *Not*I sites that are methylated in a normal genome and, therefore, do not contribute to the normal RLGS pattern. Currently, we are unable to predict which *Not*I sites are normally methylated. Despite these imperfections, we were successful in correctly predicting the sequence of these four spots on the first try. Efforts are underway to refine the DNA fragment

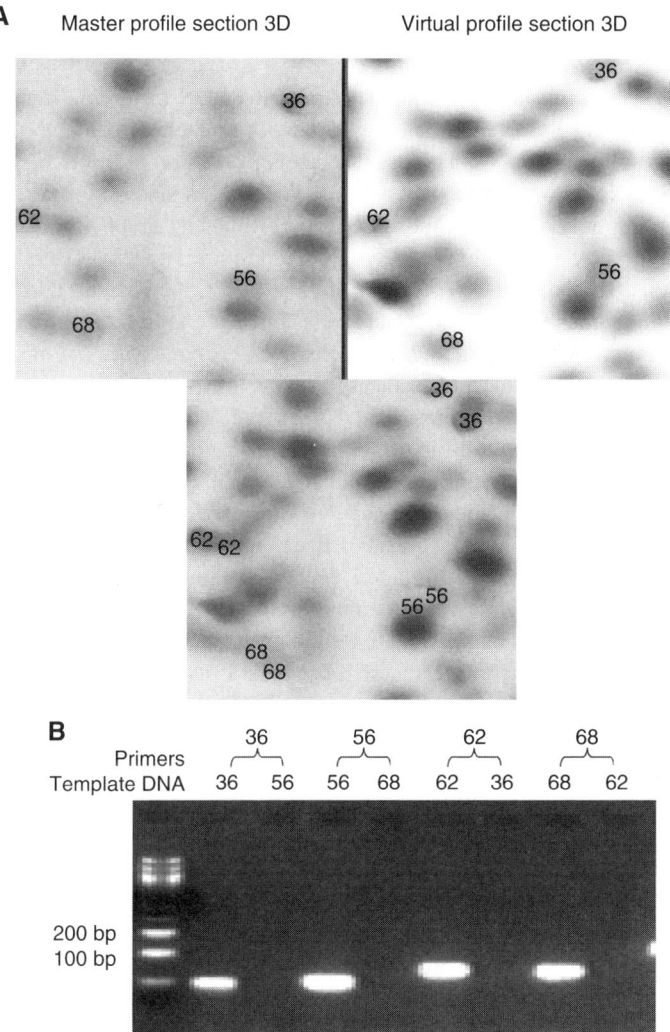

FIGURE 8.4 Spot sequence identification by vRLGS: (**A**) The left panel shows a portion of section 3D of the master RLGS profile and the right panel shows the same region from the vRLGS profile. The lower panel is a composite of the two images. Spots 3D36, 3D56, 3D62, and 3D68 are indicated on the master profile, and the corresponding candidate virtual spots are indicated on the virtual profile. The virtual spots are linked to sequence information used to design PCR primers for the confirmation the matching real and virtual fragments. (**B**) PCR primers were designed for each candidate sequence and used to amplify spot DNA eluted from the RLGS gel, or from one of the other spots as a negative control. In each case, a PCR product is produced only when the predicted spot DNA is used as template, indicating that the correct sequence was predicted for the spots using the vRLGS.

migration modeling and prediction of which *Not*I sites (by sequence context) should be normally methylated and, therefore, would not contribute to the pattern.

INCREASING GENOME COVERAGE BY SELECTING DIFFERENT ENZYME COMBINATIONS

In addition to the myriad applications of virtual RLGS in identifying RLGS spot sequences, virtual RLGS can also be used as an *in silico* experimental tool to increase the coverage of CpG islands. The optimal range of resolution in the first dimension of RLGS gels is between 5 kb and 500 bp. Within this size range, RLGS analysis is interrogating only a small proportion of the estimated 29,000 CpG islands in the human genome by using the *Not*I–*Eco*RV–*Hin*fI and *Asc*I–*Eco*RV–*Hin*fI enzyme combinations. In order to increase our coverage of CpG islands, it is possible to change the enzymes used in the RLGS gels. First, we could change the landmark enzyme to another GC-rich methylation-sensitive enzyme that cuts primarily in CpG islands. Second, we could change the second or third enzymes so that a different subset of CpG islands harboring the landmark site migrates into the window of optimal gel resolution. The difficulty comes in determining what new enzymes to use. The considerations should be that the new enzyme combinations, regardless of whether it is a change in the landmark enzyme or the second or third enzymes, should produce a pattern of spots at a density that allows adequate resolution. Additionally, the CpG islands displayed should have minimal overlap with the CpG islands displayed for other enzyme combinations.

Using vRLGS, it will be possible to test an unlimited number of possible enzyme combinations to define a subset that will produce a pattern of spots of appropriate density. This subset will be further reduced to a working set of enzyme combinations by determining which set of CpG islands is queried with each combination and by identifying the optimal set of combinations to maximize CpG island coverage in the minimum number of profiles. This will greatly enhance the utility of RLGS as a gene discovery tool and allow for a more complete picture of CpG island hyper-methylation.

SUMMARY AND FUTURE

The study of aberrant methylation has provided important insights into the patho-genesis of cancer and other diseases. Some ways in which RLGS contributed to this knowledge have been described in this chapter. When RLGS was first introduced in 1991, the data were limited to descriptive reports on the numbers of spots lost or gained on a profile.[5] For many years, it was not possible to link this data with genomic sequence, and mechanistic investigations were hampered. The subsequent development of direct cloning strategies,[63] arrayed plasmid libraries,[64,65] and com-puter-based analysis tools[10,28,68] has enabled investigators to focus on specific genes and confirmed that the number of methylated CpG islands in tumors was indeed underestimated by previous approaches.[26] Furthermore, it was determined that some methylation events occur preferentially in certain tumor types, lending credence to the hypothesis that some tumors harbor a characteristic methylation signature.[24]

The ability to associate specific genes with RLGS loci has opened many prom-ising avenues. The identification of novel cancer-related genes such as WIT1[56] and BMP3B[55] provides a starting point for *in vitro* and *in vivo* experiments designed to investigate possible tumor suppressor function. The construction of a panel of methylated sequences detected by RLGS will allow screening of large numbers of tumors in the hopes of deciphering methylation signatures in various malignancies. Such panels can also be used for in-depth analysis of luminal fluids to validate these targets as markers for cancer screening and early detection of relapse, as well as therapeutic monitoring in patients receiving pharmacologic demethylating agents. The recent development of arrayed rat[70] and mouse (Plass, unpublished data) *Not*I–*Eco*RV cloning libraries will greatly enhance the value of RLGS in the investigation of laboratory animals.

It is clear that aberrant DNA methylation is a key factor in the pathogenesis of cancer and other diseases. In order to fully understand the mechanisms by which epigenetic alterations participate in normal and neoplastic processes, we need to continue to explore various model systems as well as primary patient samples. Data obtained through genome scanning methods such as RLGS provide the foundation for further hypothesis-driven research to address these very important and basic questions.

REFERENCES

1. Herman, J. G. and Baylin, S. B., Gene silencing in cancer in association with promoter hypermethylation. *N. Engl. J. Med.,* 349, 2042–2054 (2003).
2. Palmisano, W. A. et al., Predicting lung cancer by detecting aberrant promoter methylation in sputum. *Cancer Res.,* 60, 5954–5958 (2000).
3. House, M. G. et al., Tumor suppressor gene hypermethylation as a predictor of gastric stromal tumor behavior. *J. Gastrointest. Surg.,* 7, 1004–1014 (2003).
4. Costello, J. F. and Plass, C., Methylation matters. *J. Med. Genet.,* 38, 285–303 (2001).
5. Hatada, I., Hayashizaki, Y., Hirotsune, S., Komatsubara, H., and Mukai, T., A genomic scanning method for higher organisms using restriction sites as landmarks. *Proc. Natl. Acad. Sci. USA,* 88, 9523–9537 (1991).
6. Plass, C. et al., Identification of Grf1 on mouse chromosome 9 as an imprinted gene by RLGS-M. *Nat. Genet.,* 14, 106–169 (1996).
7. Rush, L. J. et al., Comprehensive cytogenetic and molecular genetic characterization of the TI-1 acute myeloid leukemia cell line reveals cross-contamination with K-562 cell line. *Blood,* 99, 1874–1876. (2002).
8. Wu, R. et al., Amplification and overexpression of the L-MYC proto-oncogene in ovarian carcinomas. *Am. J. Pathol.,* 162, 1603–1610 (2003).
9. Ohsumi, T. et al., Loss of heterozygosity in chromosomes 1, 5, 7 and 13 in mouse hepatoma detected by systematic genome-wide scanning using RLGS genetic map. *Biochem. Biophys. Res. Commun.,* 212, 632–639 (1995).
10. Zardo, G. et al., Integrated genomic and epigenomic analyses pinpoint biallelic gene inactivation in tumors. *Nat. Genet.,* 32, 453–458 (2002).
11. Nagai, H. et al., Isolation of NotI clusters hypomethylated in HBV-integrated hepatocellular carcinomas by two-dimensional electrophoresis. *DNA Res.,* 6, 219–225 (1999).
12. Dai, Z. et al., An AscI Boundary Library for the Studies of Genetic and Epigenetic Alterations in CpG Islands. *Genome Res.,* 12, 1591–1598 (2002).

13. Plass, C. and Soloway, P. D., DNA methylation, imprinting and cancer. *Eur. J. Hum. Genet.*, 10, 6–16 (2002).

14. Nicholls, R. D., Saitoh, S., and Horsthemke, B., Imprinting in Prader-Willi and Angelman syndromes. *Trends Genet.*, 14, 194–200 (1998).

15. Reik, W. and Maher, E. R., Imprinting in clusters: lessons from Beckwith-Wiedemann syndrome. *Trends Genet.*, 13, 330–334 (1997).

16. Jirtle, R. L., Genomic imprinting and cancer. *Exp. Cell Res.*, 248, 18–24 (1999).

17. Li, E., Beard, C., and Jaenisch, R., Role for DNA methylation in genomic imprinting. *Nature*, 366, 362–365 (1993).

18. Okazaki, Y. et al., Direct Detection and Isolation of Restriction Landmark Genomic Scanning (Rlgs) Spot Dna Markers Tightly Linked to a Specific Trait By Using the Rlgs Spot-Bombing Method. *Proc. Natl. Acad. Sci. USA*, 92, 5610–5614 (1995).

19. Shibata, H. et al., Genetic mapping and systematic screening of mouse endogenously imprinted loci detected with restriction landmark genome scanning method (RLGS). *Mamm. Genome*, 5, 797–800 (1994).

20. Shibata, H. et al., The use of restriction landmark genomic scanning to scan the mouse genome for endogenous loci with imprinted patterns of methylation. *Electrophoresis*, 16, 210–217 (1995).

21. Hayashizaki, Y. et al., Identification of an imprinted U2af binding protein related sequence on mouse chromosome 11 using the RLGS method. *Nat. Genet.*, 6, 33–40 (1994).

22. Pearsall, R. S. et al., Absence of imprinting in U2AFBPL, a human homologue of the imprinted mouse gene U2afbp-rs. *Biochem. Biophys. Res. Commun.*, 222, 171–177 (1996).

23. Kamiya, M. et al., The cell cycle control gene ZAC/PLAGL1 is imprinted—a strong candidate gene for transient neonatal diabetes. *Hum. Mol. Genet.*, 9, 453–460 (2000).

24. Costello, J. F. et al., Aberrant CpG-island methylation has non-random and tumour-type-specific patterns. *Natl. Genet.*, 24, 132–138. (2000).

25. Smiraglia, D. J. et al., Differential targets of CpG island hypermethylation in primary and metastatic head and neck squamous cell carcinoma (HNSCC). *J. Med. Genet.*, 40, 25–33 (2003).

26. Rush, L. J. et al., Novel methylation targets in de novo acute myeloid leukemia with prevalence of chromosome 11 loci. *Blood*, 97, 3226–3233 (2001).

27. Costello, J. F., Plass, C., and Cavenee, W. K., Aberrant methylation of genes in low-grade astrocytomas. *Brain Tumor Pathol.*, 17, 49–56 (2000).

28. Hong, C., Bollen, A. W., and Costello, J. F., The contribution of genetic and epigenetic mechanisms to gene silencing in oligodendrogliomas. *Cancer Res.*, 63, 7600–7605 (2003).

29. Nakamura, M. et al., Genomic alterations of human gliomas detected by restriction landmark genomic scanning. *Brain Tumor Pathol.*, 14, 13–7 (1997).

30. Costello, J. F. et al., Cyclin-dependent kinase 6 (CDK6) amplification in human gliomas identified using two-dimensional separation of genomic DNA. *Cancer Res.*, 57, 1250–1254 (1997).

31. Nagai, H. et al., Genomic analysis of human hepatocellular carcinomas using Restriction Landmark Genomic Scanning. *Cancer Detect. Prev.*, 17, 399–404 (1993).

32. Nagai, H. et al., Aberration of genomic DNA in association with human hepatocellular carcinomas detected by 2-dimensional gel analysis. *Cancer Res.*, 54, 1545–1550 (1994).

33. Nagai, H., Tsumura, H., Ponglikitmongkol, M., Kim, Y. S., and Matsubara, K., Genomic aberrations in human hepatoblastomas detected by 2-dimensional gel analysis. *Cancer Res.,* 55, 4549–4551 (1995).

34. Nagai, H. et al., Genomic aberrations in early stage human hepatocellular carcinomas. *Cancer* 82, 454–461 (1998).

35. Itano, O. et al., A new predictive factor for hepatocellular carcinoma based on two-dimensional electrophoresis of genomic DNA. *Oncogene,* 19, 1676–1683 (2000).

36. Itano, O. et al., Correlation of postoperative recurrence in hepatocellular carcinoma with demethylation of repetitive sequences. *Oncogene,* 21, 789–797 (2002).

37. Dai, Z. Y. et al., Global methylation profiling of lung cancer identifies novel methylated genes. *Neoplasia,* 3, 314–323 (2001).

38. Dai, Z. et al., A comprehensive search for DNA amplification in lung cancer identifies inhibitors of apoptosis cIAP1 and cIAP2 as candidate oncogenes. *Hum. Mol. Genet.,* 12, 791–801 (2003).

39. Inui, T. et al., Genomic alterations in human meningiomas detected by restriction landmark genomic scanning and immunohistochemical studies. *Int. J. Oncol.,* 15, 459–466 (1999).

40. Yamamoto, K. et al., DNA alterations in human oral squamous cell carcinomas detected by restriction landmark genomic scanning. *J. Oral. Pathol. Med.,* 28, 102–106 (1999).

41. Thoraval, D. et al., Demethylation of repetitive DNA sequences in neuroblastoma. *Genes Chromo. Cancer,* 17, 234–244 (1996).

42. Kuick, R. et al., Studies of the inheritance of human ribosomal DNA variants detected in two-dimensional separations of genomic restriction fragments. *Genetics,* 144, 307–316. (1996).

43. Smiraglia, D. J. et al., Distinct epigenetic phenotypes in seminomatous and nonseminomatous testicular germ cell tumors. *Oncogene,* 21, 3909–3916. (2002).

44. Fruhwald, M. C. et al., Aberrant promoter methylation of previously unidentified target genes is a common abnormality in medulloblastomas—implications for tumor biology and potential clinical utility. *Oncogene,* 20, 5033–5042. (2001).

45. Miwa, W., Yashima, K., Sekine, T., and Sekiya, T., Demethylation of a repetitive DNA sequence in human cancers. *Electrophoresis,* 16, 227–232 (1995).

46. Akama, T. O. et al., Restriction landmark genomic scanning (RLGS-M)-based genome-wide scanning of mouse liver tumors for alterations in DNA methylation status. *Cancer Res.,* 57, 3294–3299 (1997).

47. Hirotsune, S. et al., New approach for detection of amplification in cancer DNA using restriction landmark genomic scanning. *Cancer Res.,* 52, 3642–3647 (1992).

48. Hughes, S. J. et al., A novel amplicon at 8p22-23 results in overexpression of cathepsin B in esophageal adenocarcinoma. *Proc. Natl. Acad. Sci. USA,* 95, 12410–12415 (1998).

49. Wimmer, K. et al., Co-amplification of a novel gene, NAG, with the N-myc gene in neuroblastoma. *Oncogene,* 18, 233–238 (1999).

50. Almeida, A. et al., GAC1, a new member of the leucine-rich repeat superfamily on chromosome band 1q32.1, is amplified and overexpressed in malignant gliomas. *Oncogene,* 16, 2997-3002 (1998).

51. Curtis, L. J. et al., Amplification of DNA sequences from chromosome 19q13.1 in human pancreatic cell lines. *Genomics,* 53, 42–55 (1998).

52. Feltus, F. A., Lee, E. K., Costello, J. F., Plass, C., and Vertino, P. M., Predicting aberrant CpG island methylation. *Proc. Natl. Acad. Sci. USA,* 100, 12253–12258 (2003).

53. Yoshikawa, H. et al., SOCS-1, a negative regulator of the JAK/STAT pathway, is silenced by methylation in human hepatocellular carcinoma and shows growth-suppression activity. *Nat. Genet.*, 28, 29–35 (2001).

54. Li, H. et al., SLC5A8, a sodium transporter, is a tumor suppressor gene silenced by methylation in human colon aberrant crypt foci and cancers. *Proc. Natl. Acad. Sci. USA*, 100, 8412–8417 (2003).

55. Dai, Z. et al., Bone morphogenetic protein 3B (BMP3B) silencing in non-small cell lung cancer. *Oncogene,* 29, 3521–3529 (2004).

56. Plass, C. et al., Restriction landmark genome scanning for aberrant methylation in primary refractory and relapsed acute myeloid leukemia; involvement of the WIT-1 gene. *Oncogene*, 18, 3159–3165 (1999).

57. Smiraglia, D. J. et al., Excessive CpG island hypermethylation in cancer cell lines versus primary human malignancies. *Hum. Mol. Genet.*, 10, 1413–1419 (2001).

58. Fruhwald, M. C. et al., Gene amplification in PNETs/medulloblastomas: mapping of a novel amplified gene within the MYCN amplicon. *J. Med. Genet.*, 37, 501–509 (2000).

59. Komatsu, S. et al., Methylation and downregulated expression of mac25/insulin-like growth factor binding protein-7 is associated with liver tumorigenesis in SV40T/t antigen transgenic mice, screened by restriction landmark genomic scanning for methylation (RLGS-M). *Biochem. Biophys. Res. Commun.*, 267, 109–117 (2000).

60. Tateno, M. et al., Identification of a novel member of the snail/Gfi-1 repressor family, mlt 1, which is methylated and silenced in liver tumors of SV40 T antigen transgenic mice. *Cancer Res.*, 61, 1144–1153 (2001).

61. Hirotsune, S. et al., Molecular cloning of polymorphic markers on RLGS gel using the spot target cloning method. *Biochem. Biophys. Res. Commun.*, 194, 1406–1412 (1993).

62. Suzuki, H., Kawai, J., Taga, C., Ozawa, N., and Watanabe, S. A, PCR-mediated method for cloning spot DNA on restriction landmark genomic scanning (RLGS) gel. *DNA Res.*, 1, 245–250 (1994).

63. Ohsumi, T. et al., A spot cloning method for restriction landmark genomic scanning. *Electrophoresis*, 16, 203–209 (1995).

64. Smiraglia, D. J. et al., A new tool for the rapid cloning of amplified and hypermethylated human DNA sequences from restriction landmark genome scanning gels. *Genomics*, 58, 254–262 (1999).

65. Plass, C. et al., An arrayed human not I-EcoRV boundary library as a tool for RLGS spot analysis. *DNA Res.*, 4, 253–255 (1997).

66. Fruhwald, M. C. et al., Aberrant hypermethylation of the major breakpoint cluster region in 17p11.2 in medulloblastomas but not supratentorial PNETs. *Genes Chromo. Cancer,* 30, 38–47. (2001).

67. Ying, A. K. et al., Methylation of the estrogen receptor-alpha gene promoter is selectively increased in proliferating human aortic smooth muscle cells. *Cardiovasc. Res.*, 46, 172–179. (2000).

68. Rouillard, J. M. et al., Virtual genome scan: a tool for restriction landmark-based scanning of the human genome. *Genome Res.*, 11, 1453–1459. (2001).

69. Yoshikawa, H. et al., Chromosomal assignment of human genomic NotI restriction fragments in a two-dimensional electrophoresis profile. *Genomics,* 31, 28–35. (1996).

70. Motiwala, T. et al., Suppression of the protein tyrosine phosphatase receptor type O gene (PTPRO) by methylation in hepatocellular carcinomas. *Oncogene*, 22, 6319–6331 (2003).

9 Quantitative Determination of 5-Methylcytosine DNA Content: HPCE and HPLC

Mario F. Fraga and Manel Esteller

CONTENTS

Methylation of cytosines in the 5 position of the pyrimidinic ring is the most important epigenetic modification in eukaryotes. In animals, methylcytosine (mC) is mainly found in cytosine–guanine (CpG) dinucleotides [1], whereas in plants it is more frequently located in cytosine–any base–guanine (CpNpGp) trinucleotide sequences (reviewed by Finnegan et al. [2]).

The most studied epigenetic alteration in mammalians is the promoter hypermethylation of CpG island-associated genes, since this process is associated with gene silencing, and it has been demonstrated to be one of the causes of the repression of tumor suppressor genes in cancer [3]. Paradoxically, cancer is also associated with the overall hypomethylation of the genomic DNA [4–6]. The explanation lays in the fact that CpG islands represent only 2% of the whole genome; thus alterations at the promoter level are masked by the global changes. The opposed behavior of the DNA methylation status of promoters and the rest of the genome is enough evidence to consider both processes as independent mechanisms and to give the deserved relevance to the global DNA methylation of the DNA.

The level of global DNA methylation has been calculated for 25 years [7]. In the beginning, it was used as a descriptive tool or marker of biological processes, such as aging and cancer [8,9] in animals and morphogenic potential in plants [10, 11]. Lately, several papers have attempted to elucidate specific mechanistic implications of the alterations of the global DNA methylation in the cell operational system. In brief, they proposed that global DNA methylation gives an extra advantage to cancer cells by increasing the genomic instability. Some of the evidence that supports this hypothesis are the embryonic stem (ES) DNMT1 knockouts that show deletions at HPRT and thymine kinase transgene [12], the ICF patients with no DNMT3b function that show numerous chromosome aberations [13], the colorectal somatic knockouts of DNMT1 that display punctual chromosomal alterations [14], and mice carrying a hypomorphic DNA methyltransferase 1 (Dnmt1) allele, which reduces Dnmt1 expression to 10% of wild-type levels resulting in substantial genome-wide hypomethylation in all tissues and high frequency of chromosome 15 trisomy [15].

From a methodological standpoint, levels of methylcytosine occurrence in the genomic DNA can be measured by high-performance separation techniques or by enzymatic–chemical means. The latter are never as sensitive as the former, and sometimes their resolution is restricted to endonuclease cleavage sites. Despite the drawbacks, enzymatic–chemical approaches are still commonly used since, unlike separation techniques, they do not require expensive and complex equipment, which is not always available. When separation devices are available, high-performance capillary electrophoresis (HPCE) may be the best choice since it is faster, cheaper, and more sensitive than high-performance liquid chromatography (HPLC) [16]. Even though *in situ* hybridization methods for studying cytosine methylation status sometimes give accurate measures of the degree of total DNA methylation, one of the most interesting aspects of these approaches is that they provide information on tissue-specific methylation patterns. By means of labeled anti-methylcytosine antibodies, DNA methylation can be monitored in metaphase chromosomes, hetero–euchromatin and, most importantly, on a cell-by-cell basis within the same sample. The latter alternative, which generally yields qualitative results, is of great interest in cancer research as it can reveal methylation differences between normal and tumor tissues in the same sample.

QUANTIFICATION OF RELATIVE mC LEVELS IN GENOMIC DNA BY HPLC

In 1977, Riggs and colleagues reported for the first time the quantification of the global relative levels of mC in genomic DNA using high-performance liquid chromatography [7]. In this pioneer approach, single bases were obtained by chemical hydrolysis in 88% formic acid at 180°C for 20 min as it had been previously described by [17]. Subsequently, the hydrolyzed products were fractioned and quantified by HPLC technologies using a Partisil SCX K218 column and 45 mM ammonium phosphate as an elution buffer. The relative amount of mC was calculated through calibration curves for the relative C and mC areas obtaining high reproducibility and sensitivity responses (only 5 to 10 μg of DNA was needed). With this innovative assay, the authors were able to reveal absolute values for overall mC levels in mammalians that are in accordance with the levels currently obtained with stronger approaches. Ten years later, Catania et al. [18] suggested the use of hydrofluoric acid instead of formic acid for chemical hydrolysis of DNA to prevent deamination of cytosine and methylcytosine, which often occurs with formic acid.

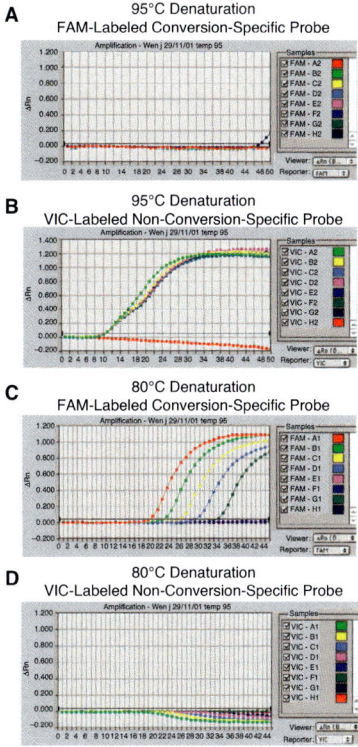

FIGURE 4.2 See black and white figure on page 59 for caption.

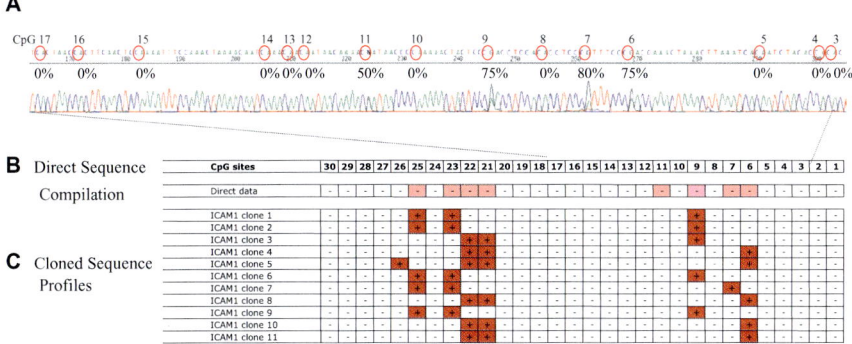

FIGURE 4.3 See black and white figure on page 61 for caption.

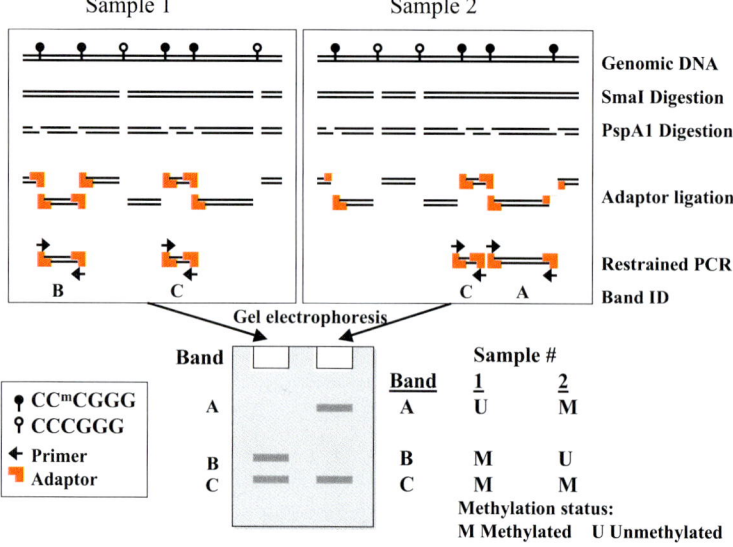

Amplification of InterMethylated Sequences (AIMS)

FIGURE 7.3   Diagram of the analysis of differential DNA methylation by amplification of intermethylated sites.

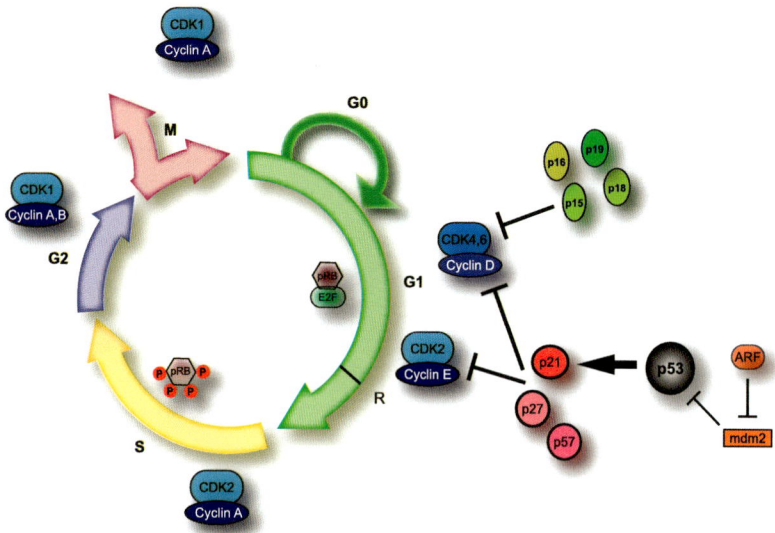

FIGURE 14.1 Schematic diagram of the cell division cycle in human cells. During G1 phase, cells increase cell size and synthesize new proteins necessary for further progression through the cell cycle. Cells can either progress through G1 and enter S-phase, or withdraw from the cell cycle (G0). During S-phase, the complete genome is duplicated once. In G2, DNA replication is completed and cells prepare to enter mitosis (M phase). During M phase, a bipolar mitotic spindle is formed, ensuring equal chromosome segregation. CDK2, CDK4, and CDK6 in combination with cyclin D and E, drive the cell cycle progression. The CDK inhibitors (INK4 and CIP/KIP) regulate and can also block their cell cycle promoting activities.

In any case, enzymatic hydrolysis of DNA is reported to be a better alternative for quantifying the degree of DNA methylation [19]. 2-Deoxymononucleosides are obtained using Nuclease PI, DNase I, and snake venom phosphodiesterase [20] or micrococcal nuclease and phosphodiesterase II [21] followed by hydrolysis with alkaline phosphatase. Resulting deoxyribonucleosides are subsequently separated by HPLC, and methylcytosine levels are quantified by comparing the relative absorbance of cytosine and methylcytosine at 254 nm in the sample with external standards of known bases. This method generally requires only 2.5 µg DNA to quantify 5-methylcytosine with a low standard deviation for replicate samples. Using these assays in the 1980s, Ehrlich and colleagues established the basis of the behavior of global mC contents in animals [22]. Other groups also provided valuable information within the field of plants [23].

One of the major problems classically associated with quantifying the overall mC levels in the genomic DNA is the sensitivity of the method because it compares the amount of one specie (mC) with the amount of another specie (C) that is 100-fold concentrated. For years, this has forced researchers to analyze large amounts of genomic DNA, a great inconvenience in regard to human primary tumors. Sensitivity of the system can be increased with mass spectrometry detection, which has a detection limit 10^6 times the limit of absorption spectroscopy detectors. Annan et al. [24] used high-performance liquid chromatography/mass spectrometry coupled with a moving-belt interface to analyze electrophoretic derivatives of methylcytosine and 5-hydroxymethyluracil nucleobases. As little as 9.9 pg (signal-to-noise ratio 5) and 180 fg (signal-to-noise ratio 10) of the respective nucleobases were detected in the negative electron-capture chemical ionization mode, and linear responses were observed over a moderate dynamic range. Alternatively, Gowher et al. [25] developed a highly specific and sensitive assay for detecting trace amounts of mC in genomic DNA. They degraded DNA to nucleosides, purified mC by HPLC and, for detection by one-dimensional and two-dimensional thin-layer chromatography, radiolabeled using deoxynucleoside kinase and [gamma-(32)P]ATP. Using this assay, they showed that mC occurs in the DNA of *D. melanogaster* at a level of approximately 1 in 1000 to 2000 cytosine residues in adult flies. This labeling strategy has already been used to quantify mC concentration [21] but, in this case, it was employed in conjunction with one-dimensional, high-performance, thin-layer chromatography using alkylamino-modified silica plates (HPTLC-NH2) instead of HPLC technology.

HPCE-BASED METHODS

The development of capillary electrophoretic techniques has given rise to a research approach that has several advantages over other current methodologies used to quantify the extent of DNA methylation. Capillary electrophoresis (CE) is a family of related separation techniques that use narrow-bore fused-silica capillaries to separate a complex array of large and small molecules (Figure 9.1A). High voltages are used to separate molecules based on differences in charge, size, and hydrophobicity. Injection into the capillary is accomplished by immersing the end of the capillary into a sample vial and applying pressure, vacuum, or voltage. Depending

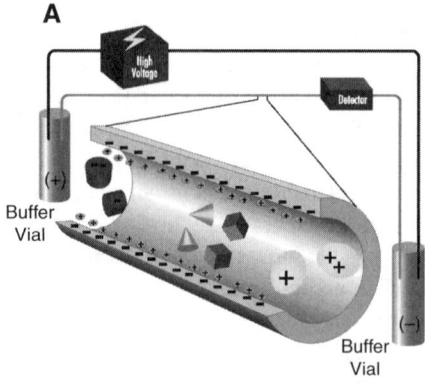

FIGURE 9.1 (**A**) Block diagram of basic CE system. The magnified view depicts the capillary interior with molecules of different size and charge separating. As illustrated, in free-solution capillary electrophoresis, the positive molecules migrate toward the negative electrode while the electroosmotic flow carries the neutral and negatively charged molecules in the same direction. (**B**) Beckman Coulter CE device.

on the types of capillary and buffers, CE can be segmented into several separation techniques of the sample components between initial and terminal electrolytes.

From its beginnings, CE has proved extremely helpful in separating various DNA components, including a number of base adducts [26], and recently Fraga et al. [16] have reported the quantification of the relative mC content of the genomic DNA of plants and animals using an HPCE system from Beckman Coulter (Figure 9.1B) to analyze acid-hydrolyzed genomic DNA. In this context, separation and quantification of cytosine and methylcytosine is only possible by the use of a sodium dodecyl sulfate (SDS) micelle system. This method is faster than HPLC (taking less than 10 min per sample). It is also reasonably inexpensive since it does not require continuous running buffers, and it displays a great potential for fractionation (theoretically, up to 10^6 plates). Biotechnology companies, such as Oncomethylome Sciences, are offering HPCE as a service to measure total 5-methylcytosine genomic content. Nevertheless, no or almost no preparative analyses are possible with HPCE systems because of the low injection volumes involved.

Release of bases from DNA by chemical means involves the production of a complicated mixture of molecules that makes detection and quantification of mC

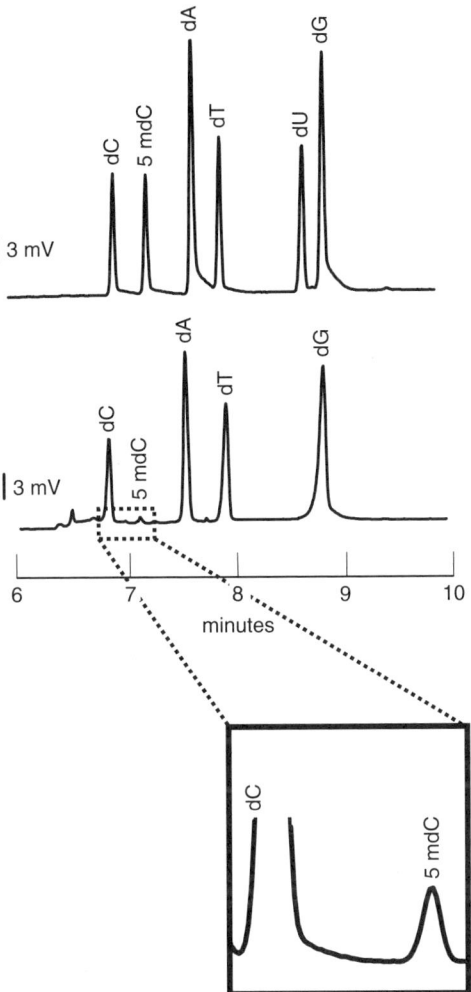

FIGURE 9.2 Quantification of global DNA methylation by HPCE; *upper panel*, a representative electropherogram for standard nucleosides (*dC*, *5mdC*, *dA*, 2-deoxythymidine, and 2-deoxyguanosine) dissolved (5 mM) in Milli-Q grade water; *lower panel*, an electropherogram obtained after enzymatic hydrolysis of 4 µg of genomic DNA from the MCF-7 human cancer cell line. A zoom of the 5-methylcytosine peak is shown in the lower square.

difficult when DNA is not sufficiently purified and concentrated. This problem can be avoided by enzymatic hydrolysis with nuclease P1 and alkaline phosphatase to produce 2-deoxymononucleosides, which are then fractionated using a modification of the previously described HPCE method [27]. Approximately 1 methylcytosine in 200 cytosine residues can be detected by this method using 1 µg genomic DNA (Figure 9.2). This is now the most commonly used methodology to quantify the global DNA methylation levels. It has been successfully used to

1. Confirm that tumors are hypomethylated in comparison to their normal counterparts [28]
2. Determine how a genomic variant for an enzyme of the S-adenosylmethionine metabolic pathway may alter the global DNA methylation levels [29]
3. Study alterations of the global DNA degree associated to PML-RAR translocations in human leukemia [30]
4. Perform a profile of global DNA methylation in human cancer cell lines and demonstrate that the treatment with the demethylating agent 5-aza-2-deoxycytidine depends on the cell type [31]
5. Ascertain the DNA demethylating properties of the anesthetic drug procaine [32]

Alternatively, laser-induced fluorescence (LIF) and mass spectrometry detectors should be used. Capillary electrophoretic separation and on-line detection by monitoring LIF have already been successfully used to quantify other modified bases in the DNA [33]. Using LIF monitoring systems, a concentration detection limit of approximately 3×10^{-10}M (91) or 1 adduct/10^7 normal nucleotides/microgram DNA [34] can be achieved. More recently, Stach et al. [35] have described a method for fluorescence detection of DNA methylation in an HPCE system. For this purpose, micrococcal nuclease DNA digestions are derivatized with a fluorescence probe, and analytes are detected in an HPCE system by LIF with an excitation wavelength of 488 nm.

In both HPLC and HPCE approaches, the absence of RNA must be ensured in order to avoid miscalculation of the relative amount of mC in the genomic DNA.

REFERENCES

1. Bird, A.P., CpG-rich islands and the function of DNA methylation, *Nature*, 321, 209, 1986.
2. Finnegan, E.J., Peacock, W.J., and Dennis, E.S., DNA methylation, a key regulator of plant development and other processes, *Curr. Opin. Genet. Dev.*, 10, 217, 2000.
3. Esteller, M., Relevance of DNA methylation in the management of cancer, *Lancet Oncol.*, 4, 351, 2003.
4. Laird, P.W. and Jaenisch, R., DNA methylation and cancer, *Hum. Mol. Genet.*, 3, 1487, 1994.
5. Esteller, M. and Herman, J.G., Cancer as an epigenetic disease: DNA methylation and chromatin alterations in human tumours, *J. Pathol.*, 196, 1, 2002.
6. Dunn, B.K., Hypomethylation: one side of a larger picture, *Ann. N.Y. Acad. Sci.*, 983, 28, 2003.
7. Singer, J., Stellwagen, R.H., Roberts-Ems, J., and Riggs, A.D., 5-Methylcytosine content of rat hepatoma DNA substituted with bromodeoxyuridine, *J. Biol. Chem.*, 252, 5509, 1997.
8. Gama-Sosa, M.A., Midgett, R.M., Slagel, V.A., Githens, S., Kuo, K.C., Gehrke, C.W., and Ehrlich, M., Tissue-specific differences in DNA methylation in various mammals, *Biochim. Biophys. Acta*, 740, 212, 1983.

9. Gama-Sosa. M.A., Slagel, V.A., Trewyn, R.W., Oxenhandler, R., Kuo, K.C., Gehrke, C.W., and Ehrlich, M., The 5-methylcytosine content of DNA from human tumors, *Nucleic Acids Res.,* 11, 6883, 1983.

10. Fraga, M.F., Canal, M.J., and Rodriguez, R., Phase-change related epigenetic and physiological changes in *Pinus radiata* D. Don., *Planta.,* 215, 672, 2002.

11. Fraga, M.F., Rodriguez, R., and Canal, M.J., Genomic DNA methylation-demethylation during aging and reinvigoration of Pinus radiata, *Tree Physiol.,* 22, 813, 2002.

12. Chen, R.Z., Pettersson, U., Beard, C., Jackson-Grusby, L., and Jaenisch, R., DNA hypomethylation leads to elevated mutation rates, *Nature*, 395, 89, 1998.

13. Xu, G.L., Bestor, T.H., Bourc'his, D., Hsieh, C.L., Tommerup, N., Bugge, M., Hulten, M., Qu, X., Russo, J.J., and Viegas-Pequignot, E., Chromosome instability and immunodeficiency syndrome caused by mutations in a DNA methyltransferase gene. *Nature,* 402, 187, 1999.

14. Rhee, I., Jair, K.W., Yen, R.W., Lengauer, C., Herman, J.G., Kinzler, K.W., Vogelstein, B., Baylin, S.B., and Schuebel, K.E., CpG methylation is maintained in human cancer cells lacking DNMT1, *Nature,* 404, 1003, 2000.

15. Gaudet, F., Hodgson, J.G., Eden, A., Jackson-Grusby, L., Dausman, J., Gray, J.W., Leonhardt, H., and Jaenisch, R., Induction of tumors in mice by genomic hypomethylation, *Science*, 300, 489, 2003.

16. Fraga, M.F., Rodriguez, R., and Canal, M.J., Rapid quantification of DNA methylation by high performance capillary electrophoresis, *Electrophoresis*, 21, 2990, 2000.

17. Wyatt, G.R., Recognition and estimation of 5-methylcytosine in nucleic acids, *Biochem. J.,* 48, 581, 1951.

18. Catania, J., Keenan, B.C., Margison, G.P., and Fairweather, D.S., Determination of 5-methylcytosine by acid hydrolysis of DNA with hydrofluoric acid, *Anal. Biochem.,* 167, 347, 1987.

19. Kuo, K.C., McCune, R.A., Gehrke, C.W., Midgett, R., and Ehrlich, M., Quantitative reversed-phase high performance liquid chromatographic determination of major and modified deoxyribonucleosides in DNA, *Nucleic Acids Res.,* 8, 4763, 1980.

20. Ehrlich, M., Gama-Sosa, M.A., Huang, L.H., Midgett, R.M., Kuo, K.C., McCune, R.A., and Gehrke, C., Amount and distribution of 5-methylcytosine in human DNA from different types of tissues of cells, *Nucleic Acids Res.,* 10, 2709, 1982.

21. Leonard, S.A., Wong, S.C., and Nyce, J.W., Quantitation of 5-methylcytosine by one-dimensional high-performance thin-layer chromatography, *J. Chromatogr.,* 1993, 645, 189, 1993.

22. Gama-Sosa, M.A., Midgett, R.M., Slagel, V.A., Githens, S., Kuo, K.C., Gehrke, C.W., and Ehrlich, M., Tissue-specific differences in DNA methylation in various mammals, *Biochim. Biophys. Acta,* 740, 212, 1983.

23. Wagner, I. and Capesius, I., Determination of 5-methylcytosine from plant DNA by high-performance liquid chromatography, *Biochim. Biophys. Acta,* 654, 52, 1981.

24. Annan, R.S., Kresbach, G.M., Giese, R.W., and Vouros, P., Trace detection of modified DNA bases via moving-belt liquid chromatography-mass spectrometry using electrophoric derivatization and negative chemical ionization, *J. Chromatogr.,* 465, 285, 1989.

25. Gowher, H., Leismann, O., and Jeltsch, A., DNA of Drosophila melanogaster contains 5-methylcytosine, *EMBO J.,* 19, 6918, 2000.

26. Norwood, C.B., Jackim, E., and Cheer, S., DNA adduct research with capillary electrophoresis, *Anal. Biochem.,* 213, 194, 1993.

27. Fraga, M.F., Uriol, E., Borja Diego, L., Berdasco, M., Esteller, M., Canal, M.J., and Rodriguez, R., High-performance capillary electrophoretic method for the quantification of 5-methyl 2-deoxycytidine in genomic DNA: application to plant, animal and human cancer tissues, *Electrophoresis,* 23, 1677, 2002.

28. Esteller, M., Fraga, M.F., Guo, M., Garcia-Foncillas, J., Hedenfalk, I., Godwin, A.K., Trojan, J., Vaurs-Barriere, C., Bignon, Y.J., Ramus, S., Benitez, J., Caldes, T., Akiyama, Y., Yuasa, Y., Launonen, V., Canal, M.J., Rodriguez, R., Capella, G., Peinado, M.A., Borg, A., Aaltonen, L.A., Ponder, B.A., Baylin, S.B., and Herman, J.G., DNA methylation patterns in hereditary human cancers mimic sporadic tumorigenesis, *Hum. Mol. Genet.,* 10, 3001, 2001.

29. Paz, M.F., Avila, S., Fraga, M.F., Pollan, M., Capella, G., Peinado, M.A., Sanchez-Cespedes, M., Herman, J.G., and Esteller, M., Germ-line variants in methyl-group metabolism genes and susceptibility to DNA methylation in normal tissues and human primary tumors, *Cancer Res.,* 62, 4519, 2002.

30. Esteller, M., Fraga, M.F., Paz, M.F., Campo, E., Colomer, D., Novo, F.J., Calasanz, M.J., Galm, O., Guo, M., Benitez, J., and Herman, J.G., Cancer epigenetics and methylation, *Science,* 297, 1807, 2002.

31. Paz, M.F., Fraga, M.F., Avila, S., Guo, M., Pollan, M., Herman, J.G., and Esteller, M., A systematic profile of DNA methylation in human cancer cell lines, *Cancer Res.,* 63, 1114, 2003.

32. Villar-Garea, A., Fraga, M.F., Espada, J., and Esteller, M., Procaine is a DNA-demethylating agent with growth-inhibitory effects in human cancer cells, *Cancer Res.,* 63, 4984, 2003.

33. Worth, C.C., Schmitz, O.J., Kliem, H.C., and Wiessler, M., Synthesis of fluorescently labeled alkylated DNA adduct standards and separation by capillary electrophoresis, *Electrophoresis,* 21, 2086, 2000.

34. Sharma, M., Jain, R., Ionescu, E., and Slocum, H.K., Capillary electrophoretic separation and laser-induced fluorescence detection of the major DNA adducts of cisplatin and carboplatin, *Anal. Biochem.,* 228, 307, 1995.

35. Stach, D., Schmitz, O.J., Stilgenbauer, S., Benner, A., Dohner, H., Wiessler, M., and Lyko, F., Capillary electrophoretic analysis of genomic DNA methylation levels, *Nucleic Acids Res.,* 31, E2, 2003.

10 Qualitative Determination of 5-Methylcytosine and Other Components of the DNA Methylation Machinery: Immunofluorescence and Chromatin Immunoprecipitation

Jesus Espada, Esteban Ballestar, and Manel Esteller

CONTENTS

0-8493-2050-X/05/$0.00+$1.50
© 2005 by CRC Press LLC

One key to understanding the exact contribution and physiological roles of DNA methylation depends on determining the spatial distribution of 5-methylcytosine, as well as that of all the machinery involved in DNA methylation and its downstream consequences. This spatial information can be obtained at two different levels: the overall distribution within the context of the nuclear architecture and the distribution along the genomic sequence.

The localization of 5-methylcytosine is not only directly related to the distribution of DNA methyltransferases (DNMTs), methyl-CpG binding proteins (MBDs), and associated proteins, but also to the result of the activity of some histone-modifying complexes. In fact, multiple connections between CpG methylation and histone modifications have been established during the past ten years [1,2]. For instance, several corepressor complexes that recruit histone deacetylase (HDAC) and histone methyltransferase (HMT) activities link DNA methylation to histone deacetylation and histone H3 K9 methylation.[3–5]

The existence of different post-translational modifications in histones was recognized in the 1960s, although only recently has it been proposed that these modifications constitute a sort of code that can be read by nuclear factors to be translated in different chromatin states (such as transcriptionally active). Thus, histones provide the necessary infrastructure for the correct and efficient operation of different nuclear machineries. However, the exact contributions of histone modifications to the different processes in which these nuclear factors are involved may depend on the spatial distribution of regulatory elements, the transcription factors involved, and the three-dimensional folding of DNA. The use of antibodies has served to provide direct biochemical and functional links between histone modifications and DNA methylation at precise nuclear and genomic locations. This chapter aims to review techniques that constitute the most important source of mechanistic information in this field: chromatin immunoprecipitation (ChIP) assays, which provide information about associations of nuclear factors and histone modifications at specific sequences; and immunolocalization, which informs about global distribution in the nucleus.

ANTIBODIES: THE POWER OF THE SPECIFICITY

The exquisite specificity of antigen and antibody recognition has provided one of the most useful properties of protein-based molecular biology tools. In contrast to chemical or enzymatic methods to determine differential folding or conformational changes in chromatin,[6–8] most of which provide only indirect information about the proteins associated with DNA, antibodies allow a direct identification of the proteins and their post-translational modification status at precise chromatin or nuclear locations.

During the past 10 years, the development of the techniques discussed in this chapter has coped with the appearance of a myriad of antibodies raised against different elements of the DNA methylation machinery. There are antibodies directed against 5-methylcytosine,[9] although most of the data has been obtained with antibodies raised against site-modified histone peptides and a variety of nuclear factors. The range of specificity of antibodies to investigate histone modifications varies between those that simply recognize modified amino acids (like anti-phospho-Ser) and others that recognize site-specific modifications of a particular histone. These antibodies can be used to identify and determine the precise spatial and temporal location of these modifications in the context of a sequence or in the nucleus.

Specificity is extremely important to guarantee the viability of these antibody-based techniques. Most of the techniques involving antibodies require several washes in highly stringent conditions that help the assays to be more specific. In general, polyclonal antibodies are preferred to monoclonal antibodies to avoid potential epitope masking problems. On the other hand, it is recommended to use affinity-purified antibodies rather than whole serum. It is important to characterize the antibodies before starting immunoprecipitation and immunolocalization experiments.

The use of antibodies against different elements of the chromatin and DNA methylation machinery is helpful in understanding the mechanistic links between histone modifications and nuclear factors. In this sense, antibodies against DNMTs, MBDs, HDACs, and HMTs have been particularly useful in establishing the connection between DNA methylation and histone modifications.[4,5,10]

IMMUNOLOCALIZATION

Immunolocalization, that is, the use of antibodies to spot specific epitopes in cells and tissues, is now a widespread methodology that allows a qualitative and morphological analysis of gene expression and protein function. In combination with the impressive advances that occurred in the fields of microscopy and image analysis, where confocal and deconvolution microscopy are illustrative examples, the potential and weight of immunolocalization have progressively increased in modern biology.[11,12] The foundations of this powerful methodology were established three decades ago and its basic principles, with very minor variations, are now current protocols in cell biology.[13–16] However, with continuous use in the bench, current protocols are applied so routinely that some aspects of the basic principles of immunolocalization are often forgotten. In this section, two of these aspects, namely fixation and mounting, are discussed in detail from the perspective of cultured cell immunolabeling. Both of them are relevant for general immunolocalization analysis and, due to the particular nature of the nuclear compartment, they also have a special importance in the study of nuclear antigens and components of the methylation machinery.

FIXATION

When facing the problem of immunolocalization of a particular antigen, the most basic challenge is choosing the correct fixation process for the sample. This is a basic principle of immunolocalization that is often ignored and erroneously resolved when applying the most current protocols with no further consideration. On one side, there exist at least eight different fixation protocols that are useful for immunolocalization analysis, but only two of them are currently used (Table 10.1). On the other side, it is well known that different fixation protocols render different immunolocalization patterns of the same antigen.[17] The object of fixation is to preserve cells and extracellular materials in such a way that there has been as little

---

TABLE 10.1
Fixation Protocols Suitable for
Immunolocalization in
Cultured Cells

Coagulant Fixatives

Cold methanol
Cold methanol/acetone
Cold methanol/acetic acid

Noncoagulant Fixatives

Neutral buffered formalin
Formaldehyde with glutaraldehyde
Formalincalcium

Coagulant/Noncoagulant Fixatives

Neutral buffered formalin with picrate
Formalin-methanol

---

alteration as possible to the structure and chemical composition of the original living sample.[18] It is clear that perfect fixation is impossible to attain and that a certain structural distortion of the starting material is always a side effect of fixation. However, it is also obvious that in each case particular attention should be paid to the process of fixation. A standard fixation protocol is not always the correct choice for the detection of a particular antigen, and this choice will determinate the quality of the information obtained.

Biological fixation can be achieved by both physical (heating and freezing) and chemical methods. Chemical fixation with liquid fixatives is the method of choice for most immunolocalization purposes. Liquid fixatives affect biological samples both physically and chemically, being that the rate of penetration of these substances in the living material is a determinant factor of the final impact on the sample. The principal physical changes produced by liquid fixatives are shrinkage, swelling, and hardening, and many chemical reactions altering the nature of biomolecules are implicated in fixation. (For a detailed discussion of biological fixation, see Pearse[13] and Kiernan.[18]) Based on their major chemical actions on the sample, liquid fixatives are grouped in two main categories: coagulants and noncoagulants.[18] Fixatives that cause coagulation of cytoplasmic proteins destroy or distort organelles and secretory granules, but they do not seriously disturb the supporting extracellular materials. Thus, coagulants are thought to produce a sponge-like proteinaceous reticulum inside the cells, which results in good overall morphological preservation, but also in an extensive loss of biochemical properties of the sample. Noncoagulant fixatives promote protein crosslinking and tend to convert the cytoplasm in an insoluble gel in which the organelles are well preserved. In this insoluble gel, the chemical properties of biomolecules are relatively maintained, but the morphological preservation is poor.

Common organic coagulants suitable for immunolocalization include acetone, ethanol, and methanol (Table 10.1). These chemical compounds displace water from biomolecules, thereby breaking hydrogen bonds in a denaturing reaction by which soluble proteins are coagulated while nucleic acids remain soluble in water and are not precipitated. Most lipids are also dissolved in this reaction. At low temperatures (–20°C) alcohols precipitate many proteins without denaturing them. These precipitated proteins retain their biochemical properties and remain soluble in water. Other coagulant fixatives of some use in immunolocalization are the fast penetrating acetic and picric acids (Table 10.1). Acetic acid coagulates nucleic acids but does not fix proteins. Picric acid coagulates biological samples by forming salts (picrates) with the basic groups of proteins but also causes a strong nucleic acid hydrolysis.

The most common noncoagulant fixative is formaldehyde (Table 10.1). Formaldehyde is a highly reactive gas usually found as a solid polymer (paraformaldehyde) or as a solution containing 37 to 40% (w/v) (formalin). Formalin solutions also contain methanol (10% v/v) as a stabilizer, an important detail that should be kept in mind during fixation. Formaldehyde reacts with many chemical groups of protein molecules, such as primary amines of N-terminal amino acids and lysine side-chains, guanidyl groups of arginine, sulphydryl groups of cysteine, aliphatic hydroxyl groups of serine and threonine, and amide nitrogen at peptide linkages (see Pearse[13] for detailed information). These reactions render hemiacetal-like adducts with free hydromethyl groups capable of further reactions with suitably positioned functional groups in the same or adjacent proteins. Formaldehyde also preserves most lipids, especially if fixing solutions contain calcium ions. Glutaraldehyde, another common noncoagulant fixative, is a bifunctional aldehyde with chemical properties similar to those of formaldehyde.

When analyzing nuclear components, the particular nature of the nuclear compartment and its specific response to the process of chemical fixation should be considered. The main component of the nucleus is the double strand of DNA and its associated proteins, that is, chromatin. Chromatin is packaged orderly inside the nucleus using the structural elements of the nuclear matrix. Emerging concepts indicate that a coordinate and dynamic interaction of the components of chromatin and nuclear matrix establish a definite three-dimensional arrangement of local regions inside the nucleus that is functionally relevant for cell functioning.[19,20] This arrangement should be preserved as well as possible in the process of fixation in order to perform a correct immunolocalization analysis of nuclear function, but it is obvious that any method of fixation induces changes in the physical state of chromatin. From a morphological point of view, formalin is not a good fixative for chromatin, but coagulant fixatives, such as alcohols and acetic acid, are recommended in order to preserve the native three-dimensional structure of the nucleus. However, coagulants are expected to destroy or alter the antigenecity and localization of epitopes. In fact, different fixatives result in different nuclear morphologies and in different patterns of epitope localization. Coagulant–noncoagulant fixative mixtures are strongly recommended for immunolocalization of nuclear proteins (Table 10.1). These mixtures are expected to combine the properties of both types of fixatives, reaching the equilibrium between nuclear morphology preservation and the maintenance of antigenecity (Figure 10.1).

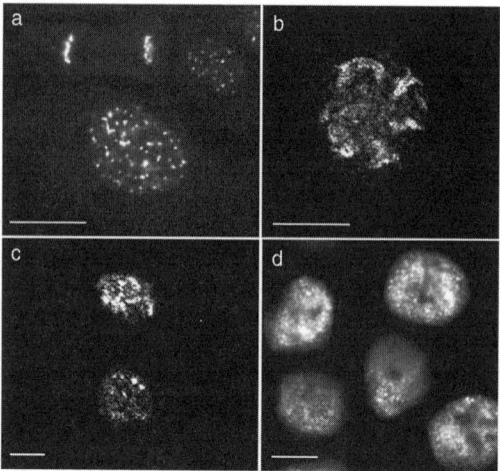

FIGURE 10.1 Fluorescence micrographs showing the localization of different components of the epigenetic machinery after immunolabeling with specific antibodies of formaldehyde/methanol fixed samples. (**a**) Distribution of 5mC in mouse Pam212 keratinocytes. Localization of two MBD proteins, MeCP2 (**b**) and MBD1 (**c**), in human MCF7 mammary carcinoma cells. (**d**) Immunolocalization pattern of acetylated histone H3 in human HCT116 colon carcinoma cells. Bars: 10 μm.

INDIRECT IMMUNOFLUORESCENCE

There exist two main approaches for immunolocalization: direct and indirect immunolabeling. In direct immunolabeling, the antibody used for antigen detection is directly coupled to a tracer molecule. Only one step is needed in order to immunolocalize the antigen. In indirect immunolabeling, a primary antibody is used for antigen recognition and a second antibody, coupled to a tracer molecule, is used to localize the primary antibody. Tracer molecules are typically fluorescent or enzymatic markers. However, due to the widespread use and superior sensitivity of fluorescence microscopy, the lower molecular weight, and easy chemical handling of fluorophores, these molecules have become the tracers of choice for immunolocalization. Current immunolocalization analysis uses mainly the indirect approach because the labeling of each primary antibody is laborious and can decrease the specificity for the antigen.[21] In addition, direct immunolocalization is less sensitive since more than one secondary antibody could theoretically bind to one molecule of primary antibody.

A feasible immunolocalization of almost any nuclear protein can be achieved by using a tested fixation protocol in combination with an indirect immunofluorescence approach, in which the availability of appropriate primary antibodies is a limiting requirement. In particular, immunolocalization of proteins implicated in the epigenetic regulation of nuclear function, such as MBDs or HDACs (Figure 10.1), have rendered much valuable information, strongly supporting results obtained by biochemical or molecular approaches.[22–24] A very special component of the epigenetic machinery, the 5-methylated cytosine residue of DNA (5mC), can also be

TABLE 10.2
Procedure for 5mC Immunolocalization in Cultured Cells

- Fix the cells in 4% formalin for at least 15 min at RT. Slightly better results are obtained fixing the cells O/N at 4°C.
- Wash a few times in dH_2O. For O/N fixing, this wash should be prolonged for at least 30 min.
- Refix the cells in cold methanol (−20°C) for 5 min.
- Wash in dH_2O and incubate the cells in 2N HCl for at least 30 min at 37°C. This is a critical step. It is expected to hydrolyze RNA thus avoiding a false 5mC signal of non-DNA species. Incubation time depends on cell type. Longer times and stronger DNA hydrolysis are good for cells with compacted nuclei. In other cells, longer incubation times may render a high background. 45 min is a good starting point.
- Wash in water and incubate the samples in Tris-Borate-EDTA (standard TBE for DNA electrophoresis) for 5 min at RT.
- Wash in water and then in PBS, block the cells in PBS-BSA, and procceed with immunostaining.

Source: Adapted from Habib, M. et al., *Exp. Cell Res.*, 249, 46, 1999.

accurately located by indirect immunolabeling (Figure 10.1). Antibodies against methylcytidine, recognizing methylcytosine residues in both DNA and RNA species, are now available.[25] The use of these antibodies for the specific location of methylated DNA requires a special protocol that includes a strong acid hydrolysis of nucleic acids (Table 10.2). Acid treatment in such specific conditions is expected to completely hydrolyze and extract RNA molecules and partially denature DNA, exposing methylcytosine residues for a correct antibody recognition.[25]

MOUNTING

The final step in immunolocalization analysis is the process of sample mounting. It is generally assumed that an aqueous medium is necessary for the preservation and examination of immunolabeled samples with fluorescent signals. Accordingly, water or glycerol–water mixtures were originally used as mounting media for immuno-fluorescent preparations. Semipermanent aqueous media containing polyvinyl alcohol (PVA) and glycerol were described as an improved alternative to glycerol–water mixtures, allowing the preservation of samples in a rigid medium.[26,27] Mowiol 40-88 and variants are now the most frequent class of PVA used for fluorescent immuno-labeling.[28,29]

Although it is generally claimed that immunofluorescence reactions are stable in Mowiol-mounted specimens stored in the dark at 4°C, common practice indicates that, in these conditions, the quality of preparations diminishes and becomes unsuitable for microscopic observation. With most antibodies, the fluorescent signal corresponding to specific immunolocalization in aqueous-mounted samples finally become diffuse and decay after variable times, ranging from days to weeks (Figure 10.2). A similar situation is observed when intercalant DNA fluorophores, such as DAPI, are used for nuclear counterstaining (Figure 10.2). A very suitable and improved alternative is permanent mounting in nonaqueous medium, such as DePeX[30] (Table 10.3). Mounting in nonaqueous medium is the procedure of choice

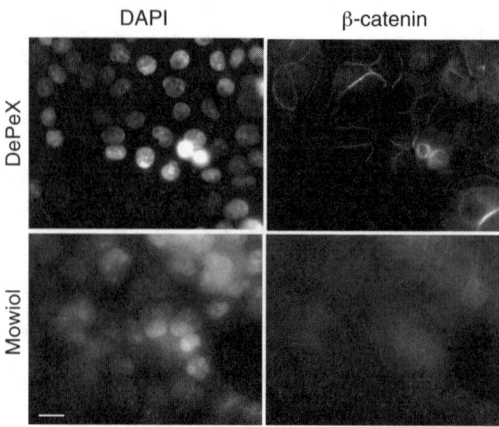

FIGURE 10.2 Fluorescence micrographs of human HCT116 colon carcinoma cells immunolabeled for β-catenin and counterstained with DAPI. Images were obtained one month after mounting in aqueous Mowiol medium (lower panels) or nonaqueous DePeX medium (upper panels). Note the diffusion and altered localization of both β-catenin and DNA fluorescent signals in aqueous Mowiol medium. Images obtained just after immunolabeling in both mounting media were quite similar to those obtained in DePeX one month after mounting. Bar: 10 μm.

TABLE 10.3
Nonaqueous Permanent Mounting of Immunofluorescence Samples

- After the last step of immunolabeling, cell cultured samples are washed two to three times in dH₂O to completely remove PBS salts
- If desired, cell nuclei are counterstained with DAPI (10 nM in 50% ethanol) for 5 min at RT.
- Wash three times in 70% ethanol and completely dehydrate the samples in 100% ethanol (each step 10 to 20 sec).
- Cell samples are finally cleared in xylene and mounted in a drop of DePeX (a nonfluorescing neutral solution of polystyrene in xylene; Serva).

Source: Adapted from Espada, J. et al., *Biotech. Histochem.*, in press.

---

for most histological and histopatholgical purposes and requires ethanol dehydration and xylol clearing of the sample.[18] It is well known that fluorescent reactions induced by a wide variety of fluorochromes and fluorescent vital probes are very well preserved after dehydration, clearing, and permanent mounting in nonaqueous media.[31–34] Absence of water and oxygen radicals in hydrophobic media prevents both emission decay and diffusion of hydrophilic fluorescent probes, almost completely eliminating background signals of medium autofluorescence. Antigen-antibody recognition is also correctly preserved after the dehydration and clearing steps required for hydrophobic mounting. Thus, nonaqueous mounting of immunofluorescent samples results in protein localization patterns that are indistinguishable from those obtained just after aqueous mounting (Figure 10.2). However, the advantage of nonaqueous over aqueous mounting of immunofluorescent samples is strongly evident a few weeks after mounting (Figure 10.2). An additional and very interesting benefit of permanent hydrophobic mounting is that immunofluorescent samples can be easily archived at room temperature and examined after months or even years.

MAPPING HISTONE MODIFICATIONS AND NUCLEAR PROTEINS BY CHROMATIN IMMUNOPRECIPITATION

ChIP assays have become very popular techniques not only to determine the modification status of histones at specific sequences, but also to identify the presence of nuclear factors at those sequences. In brief, this biochemical approach is performed through fixation of the cells with formaldehyde to prevent dissociation of DNA-associated proteins from their target DNA sequence in subsequent steps. The protocol continues by fragmentation of chromatin, by micrococcal nuclease, or by or sonication, and the cleared lysate is incubated with a primary antibody. The antibody-antigen complexes are recovered by addition of protein A or protein G Sepharose beads. The DNA bound to the antibodies can then be recovered and analyzed by PCR amplification with specific primers or hybridization with radioactive probes (Figure 10.3). Recently, a combination of ChIP assays with genome-wide techniques has broadened the potential of this technique.

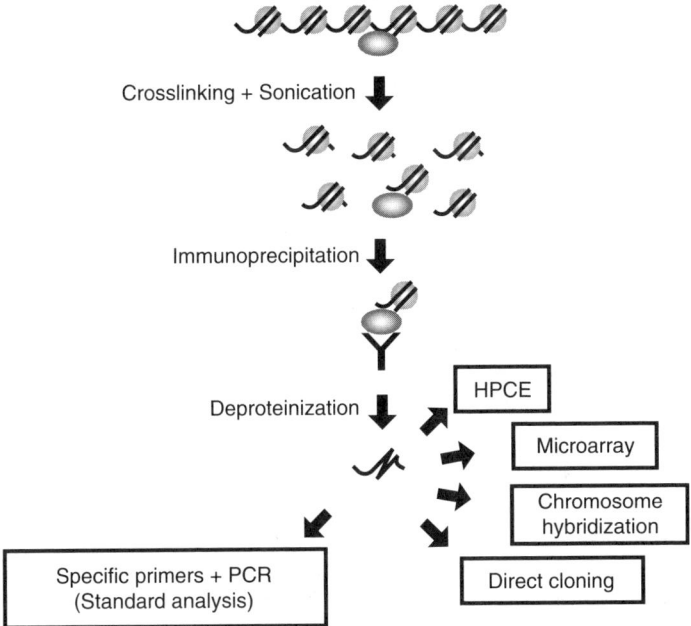

FIGURE 10.3 Schematic representation of the ChIP assays. Standard analysis (PCR amplification), (bottom left corner), or combination with genome-wide analyses (bottom right corner) are depicted.

During the past few years, excellent methodological reviews describing this technique have been written,[35-38] as the technique has become one of the most commonly used tools to investigate the profile of histone modifications and binding of transcription factors to specific sequences.[39-42]

Since there are detailed protocols describing the ChIP assay,[37,38] this section will focus on several critical steps, including optimization of the crosslinking, DNA fragmentation, and quality of the antibody, as well as on some of the applications of the ChIP assay.

A FEW TECHNICAL CONSIDERATIONS

The most common crosslinking agent used in ChIP assays is formaldehyde, a dipolar reagent that produces both protein–nucleic acid and protein–protein crosslinks through the imino groups of amino acids, such as Lys, Arg, and His, and DNA (adenines and cytosines). A key property of the crosslinks obtained by using form-aldehyde is its reversibility, which is achieved by treatment at low pH in aqueous solution or incubation at 60 to 70°C in the presence of SDS. Due to the small size of formaldehyde (2Å) only proteins located within this distance from DNA will become crosslinked. Some of the chromatin-modifying enzymes, such as histone deacetylase, do not directly bind to DNA, and their gene-specific regulatory functions occur through recruitment by DNA-binding proteins to a particular regulatory sequence. Although these proteins do not exhibit DNA-binding properties, it is possible to investigate their interaction with particular sequences by using additional protein–protein crosslinkers.[43] For instance, dimethyl adipimidate (DMA) has been used to investigate the association of the yeast HDAC Rpd3.[44]

Efficient fixation of proteins to DNA is crucial for the ChIP assay. Standard conditions for formaldehyde crosslinking usually consist of a concentration of 1% between 15 min and 1 h, depending on the proteins to be analyzed. It is important to avoid long formaldehyde crosslinking treatment as it results in an increase of resistence to fragmentation by sonication and will decrease the efficiency of the technique.

On the other hand, formaldehyde is a moderately denaturing agent for proteins and high concentration, and long exposure to this reagent may result in the loss of antigen epitopes. It is advisable to empirically determine the sensitivity of proteins to formaldehyde. After standard fixing conditions at different times, immunolocalization analysis can detect loss of fluorescence signal due to denaturation.

When choosing fixation conditions, it is important to ensure that the increased mechanical resistence of chromatin will still allow fragmentation by sonication. In fact, the size of the chromatin fragments is the second critical parameter to consider when performing ChIP assays. The size of the chromatin fragments will determine both the yield of the amount of immunoprecipitated material and the resolution of the technique. Chromatin fragmentation is generally achieved by sonication (although micrococcal nuclease can also be used in protocols that avoid fixation by formaldehyde). Chromatin fragmentation must be optimized for each sonicator and, for that, pilot experiments with different sonication conditions must be performed prior to an immunoprecipation experiment.

In many different studies, accurate mapping can be achieved by designing primers that amplify DNA fragments of around 200 bp. Large chromatin fragments are immunoprecipitated less efficiently than small fragments, at least in a specific fashion. On the other hand, the size of the fragments determines the resolution of the technique. Fragments should not largely exceed the size of the sequence to be analyzed. If the average chromatin fragments are much larger than the sequence to be PCR amplified or probed, one cannot be sure that the protein for which the antibody was used is actually bound to that particular sequence or in the neighboring regions.

Finally, the quality of the antibody is extremely important in ChIP assays. Considerations to be taken when planning a ChIP assay related to the antibodies should, in general, be similar to the details previously mentioned. First, it is essential to ensure that the antibody is efficient in pulling down the antigen and, second, that most of the material represents specific immunoprecipitated DNA, with respect to nonspecifically precipitated DNA. Generally, "no antibody" controls are included. Finally, it is advisable to include a preimmune serum, when total antiserum is used in the ChIP assay, or an immunoglobulin as additional negative controls.

ChIP Assays Are Mapping the Genome Road

The use of antibodies in ChIP assays has served to provide direct functional and mechanistic links among histone modifications, nuclear factor binding, DNA methylation, and different chromatin states. Although many early studies based on biochemical fractionation suggested that histone acetylation is linked to transcriptional activation, direct evidence was obtained when using antibodies to acetylated core histones in ChIP assays. Immunoselection of DNA fragments containing acetylated histones led to a major enrichment of transcribed sequences.[45] Chromatin immunoprecipitation has also served to establish functional connections between histone deacetylation and methylation of DNA sequences.[46-48] The possibility of using ChIP assays to investigate recruitment of different factors to specific sequences has served to detect interaction of different elements of the DNA methylation machinery to methylated sequences, like MBDs that are recruited to hypermethylated genes in cancer.[49-51] This information, together with immunoprecipitation results that are able to connect different subunits of multiprotein complexes, provide mechanistic insights between different elements of this machinery. Moreover, these types of studies not only serve to map the chromatin of genes, but also help us to understand the histone code.

The Use of ChIPs to Identify Novel Methylated Targets

The completion of the human genome has provided a huge valuable database for scientists. The sequence contains both the information corresponding to the entire proteome and part of the infrastructure involved in the ordered regulation of gene expression. We now know that regulation of gene expression depends not only on the presence of cis regulatory elements that are contained in the DNA sequence, and that chromatin structure, histone modifications, and the binding of many different transcriptional factors participate in this process. The past years have witnessed the development of novel genome-wide strategies conducted to analyze the vast amounts of information that is now available from the human genome.

One strategy that can provide organized information about how the information encoded in the genome is used arises from the possibility of isolating functionally characterized DNA sequences and identifying them in the context of the genome. For instance, if it were possible to isolate the entire collection of regulatory sequences that are bound by a particular transcription factor, and these sequences could be identified by using the information from the human genome, a global picture of the regulation mediated by this factor could be obtained.

The combination of ChIP assays, which are useful to isolate DNA sequences bound by a factor or with a particular histone modification pattern, with genome-wide strategies make possible this proposal. In this sense, two major approaches have been used to date: the first one is based on the combination of ChIP assays with direct cloning and sequencing;[52] the second one requires hybridization of the ChIP sequences on a specialized microarray (ChIP on chip).[51,53]

The first approach, direct cloning and sequencing of immunoprecipitated DNA, has been used to identify sequences that are bound by a human transcription factor.[52] The major problem in this strategy is eliminating as much nonspecifically precipitated DNA as possible. In a standard ChIP assay, a significant amount of DNA is precipitated nonspecifically. This is generally not a problem because specific primers are used. To minimize the amount of this nonspecific DNA, sequential immunoprecipitations are performed.[52]

A most promising genome-wide approach is provided by combining the ChIP assays with hybridization on a microarray (ChIP on chip). This approach has been succesfully used to investigate transcription factor targets in yeast[54,55] and human samples.[51,53] This technique relies on the availability of appropriate microarrays. Since most of the available microarrays contain cDNA sequences, using this approach to identify regulatory regions has been limited to the yeast system and the use of a CpG microarray. However, there are multiple efforts to construct appropiate genomic microarrays and soon there will be several microarrays available for these studies.

ChIP on chip analyses require an amount of DNA that largely exceeds the standard amounts of recovered DNA. For instance, standard conditions for ChIP require the use of about 1 million cells to yield between 10 and 100 ng of immunoprecipitated DNA. In ChIP on chip analysis, submicrogram amounts of DNA are required. Therefore, when planning this type of experiment, it is important to scale up the amounts of cells and antibodies to convenient levels.

ChIP on chip assays produce comprehensive information about histone modifications and nuclear factor binding. In order to produce useful information, additional experiments are necessary. First, independent ChIP assays are required to validate the identified sites. Individual ChIPs help to minimize the number of false positives. Second, functional experiments complement and also validate the conclusions obtained from the assays. For instance, when applying the ChIP on chip assays to investigate DNA sequences with particular histone modification patterns, for which a particular known transcriptional status is associated, expression studies can provide this type of functional information. The possibilities of different combinations of the ChIP assay with genome-wide technologies is of enormous potential since it will facilitate the discovery of novel targets from a functional point of view.

REFERENCES

1. Dobosy, J.R. and Selker, E.U., Emerging connections between DNA methylation and histone acetylation. *Cell Mol. Life Sci.*, 58, 721, 2001.
2. Burgers, W.A., Fuks, F., and Kouzarides, T., DNA methyltransferases get connected to chromatin. *Trends Genet.*, 18, 275, 2002.
3. Jones, P.L., Veenstra, G.J., Wade, P.A., Vermaak, D., Kass, S.U., Landsberger, N., Strouboulis, J., and Wolffe, A.P., Methylated DNA and MeCP2 recruit histone deacetylase to repress transcription. *Natl. Genet.*, 19, 187, 1998.
4. Robertson, K.D., Ait-Si-Ali, S., Yokochi, T., Wade, P.A., Jones, P.L., and Wolffe, A.P., DNMT1 forms a complex with Rb, E2F1 and HDAC1 and represses transcription from E2F-responsive promoters. *Natl. Genet.*, 25, 338, 2000.
5. Fuks, F., Hurd, P.J., Deplus, R., and Kouzarides, T., The DNA methyltransferases associate with HP1 and the SUV39H1 histone methyltransferase. *Nucleic Acids Res.*, 31, 2305, 2003.
6. McGhee, J.D., Felsenfeld, G., and Eisenberg, H., Nucleosome structure and conformational changes. *Biophys. J.*, 32, 261, 1980.
7. Gregory, P.D., Barbaric, S., and Horz, W., Restriction nucleases as probes for chromatin structure. *Methods Mol. Biol.*, 119, 417, 1999.
8. Ballestar, E., Boix-Chornet, M., and Franco, L., Conformational changes in the nucleosome followed by the selective accessibility of histone glutamines in the transglutaminase reaction: effects of ionic strength. *Biochemistry*, 40, 1922, 2001.
9. Barbin, A. et al., New sites of methylcytosine-rich DNA detected on metaphase chromosomes. *Hum. Genet.*, 94, 684, 1994.
10. Datta, J., Ghoshal, K., Sharma, S.M., Tajima, S., and Jacob, S.T., Biochemical fractionation reveals association of DNA methyltransferase (Dnmt) 3b with Dnmt1 and that of Dnmt 3a with a histone H3 methyltransferase and Hdac1. *J. Cell Biochem.*, 88, 855, 2003.
11. Polak, J.M. and Van Noorden, S., *Introduction to immunocytochemistry*, 2nd ed. Royal Microscopical Society Microscopy Handbooks, Vol. 37. BIOS Scientific Publishers, Oxford, UK, 1999.
12. Wright, S.J. and Wright, D.J., Introduction to confocal microscopy. *Meth. Cell Biol.* 70, 2, 2002.
13. Pearse, A.G.E., *Histochemistry: theoretical and applied*, 3rd ed. Churchill-Livingstone, Edinburgh, UK, 1972.
14. Sternberger, L.A., *Immunocytochemistry*, 3rd ed. Wiley, New York, 1986.
15. Barret, J.T., *Textbook of immunology: an introduction to immunocytochemistry and immunobiology*, 5th ed. Mosby, St. Louis, 1987
16. Beltz, B.S. and Burtz, G.D., *Immunocytochemical techniques: principles and practice*. Blackwell, Cambridge, MA, 1989.
17. Melan, M.A, and Sluder, G., Redistribution and differential extraction of soluble proteins in permeabilized cultured cells. Implications for immunofluorescence microscopy. *J. Cell Science*, 101, 731, 1992.

18. Kiernan, J.A., *Histological and histochemical methods. Theory and practice.* 3rd ed. Arnold Publishers, London, 1999.

19. Cremer, T. and Cremer, C., Chromosome territories, nuclear architecture and gene regulation in mammalian cells. *Nature Rev. Genet.*, 2, 292, 2001.

20. Trouche, D., Khochbin, S., and Dimitrov, S., Chromatin and epigenetics: dynamic organization meets regulated function. *Cell*, 12, 281, 2003.

21. Javois, L.C., Direct immunofluorescent labeling of cells. *Methods Mol. Biol.*, 115,107, 1999.

22. Wysocka, J., Myers, M.P., Laherty, C.D., Eisenman, R.N., and Herr, W., Human Sin3 deacetylase and trithorax-related Set1/Ash2 histone H3-K4 methyltransferase are tethered together selectively by the cell-proliferation factor HCF-1. *Genes Dev.*,17, 896, 2003.

23. Suzuki, M., Yamada, T., Kihara-Negishi, F., Sakurai, T., and Oikawa, T., Direct association between PU.1 and MeCP2 that recruits mSin3A-HDAC complex for PU.1-mediated transcriptional repression. *Oncogene*, 22, 8688, 2003.

24. Watamoto, K., Towatari, M., Ozawa, Y., Miyata, Y., Okamoto, M., Abe, A., Naoe, T., and Saito, H., Altered interaction of HDAC5 with GATA-1 during MEL cell differentiation. *Oncogene*, 22, 9176, 2003.

25. Habib, M., Fares, F., Bourgeois, C.A., Bella, C., Bernardino, J., Hernandez-Blazquez, F., de Capoa, A., and Niveleau, A. DNA global hypomethylation in EBV-transformed interphase nuclei. *Exp. Cell Res.*, 249, 46, 1999.

26. Rodriguez, J. and Deinhardt, F., Preparation of semipermanent mounting medium for fluorescent antibody studies. *Virology*, 12, 316, 1960.

27. Thomason, B.M. and Cowart, G.S., Evaluation of polyvinyl alcohols as semipermanent mountants for fluorescent-antibody studies. *J. Bacteriology*, 93, 768, 1967.

28. Osborn, M. and Weber, K., Immunofluorescence and immunocytochemical procedures with high affinity purified antibodies: tubulin-containing structures. *Meth. Cell Biol.*, 24, 97, 1982.

29. Baschong, W., Suetterlin, R., and Laeng, R.H., Control of autofluorescence of archival formaldehyde-fixed, paraffin-embedded tissue in confocal laser scanning microscopy (CLSM). *J. Histochem. Cytochem.*, 49, 1565, 2001.

30. Espada, J., Juarranz, A., Galaz, S., Cañete, M., Villanueva, A., and Stockert, J.C., Nonaqueous permanent mounting for immunofluorescence microscopy. *Biotech. Histochem*, in press.

31. Del Castillo, P., Llorente, A.R., Gómez, A., Gosálvez, J., Goyanes, V.J., and Stockert, J.C., 1. New fluorescence reactions in DNA cytochemistry. 2. Microscopic and spectroscopic studies on fluorescent aluminum complexes. *Analyt. Quant. Cytol. Histol.*, 12, 11, 1990.

32. Juarranz, A,, Villanueva, A., Cañete, M., Polo, S., Domínguez, V., and Stockert, J.C., Microscopical and spectroscopic studies on the fluorescence of a daunomycin-aluminum complex. *Histochem. J.*, 31, 201, 1999.

33. Stockert, J.C., Cytochemistry of mast cells: new fluorescent methods selective for sulfated glycosaminoglycans. *Acta. Histochem.*, 102, 259, 2000.

34. Cañete, M., Juarranz, A., López-Nieva, P., Alonso-Torcal, C., Villanueva, A., Stockert, J.C., Fixation and permanent mounting of fluorescent probes after vital labelling of cultured cells. *Acta. Histochem.*, 103, 117, 2001.

35. Crane-Robinson, C. and Wolffe, A.P., Immunological analysis of chromatin: FIS and CHIPS. *Trends Genet.*, 14, 477, 1998.

36. Orlando, V., Mapping chromosomal proteins *in vivo* by formaldehyde-crosslinked-chromatin immunoprecipitation, *Trends Biochem. Sci.*, 25, 99–104, 2000.

37. Wells, J. and Farnham, P.J., Characterizing transcription factor binding sites using formaldehyde crosslinking and immunoprecipitation. *Methods*, 26, 48, 2002.

38. Spencer, V.A., Sun, J.M., Li, L., and Davie, J.R., Chromatin immunoprecipitation: a tool for studying histone acetylation and transcription factor binding. *Methods*, 31, 67, 2003.

39. Orlando, V. and Paro, R., Mapping Polycomb-repressed domains in the bithorax complex using *in vivo* formaldehyde cross-linked chromatin. *Cell,* 75, 1187, 1993.

40. Vettese-Dadey, M., Grant, P.A., Hebbes, T.R., Crane- Robinson, C., Allis, C.D., and Workman, J.L., Acetylation of histone H4 plays a primary role in enhancing transcription factor binding to nucleosomal DNA *in vitro*. *EMBO J.*, 15, 2508, 1996.

41. Fournier, C., Goto, Y., Ballestar, E., Delaval, K., Hever, A., Esteller, M., and Feil, R., Allele-specific histone lysine methylation marks regulatory regions at imprinted mouse genes. *EMBO J.*, 21, 6560, 2002.

42. Peinado, H., Ballestar, E., Esteller, M., and Cano, A., The transcription factor Snail mediates E-cadherin repression by the recruitment of the Sin3A/histone deacetylase 1 (HDAC1)/HDAC2 Complex. *Mol. Cell Biol.*, 24, 306, 2004.

43. Kurdistani, S.K. and Grunstein, M., *In vivo* protein-protein and protein-DNA crosslinking for genomewide binding microarray. *Methods*, 31, 90, 2003.

44. Kurdistani, S.K., Robyr, D., Tavazoie, S., and Grunstein. M., Genome-wide binding map of the histone deacetylase Rpd3 in yeast. *Natl. Genet.*, 31, 248, 2002.

45. Hebbes, T.R., Thorne, A.W., and Crane-Robinson, C., A direct link between core histone acetylation and transcriptionally active chromatin. *EMBO J.*, 7, 1395, 1988.

46. Torres et al., Liver-specific methionine adenosyltransferase MAT1A gene expression is associated with a specific pattern of promoter methylation and histone acetylation: implications for MAT1A silencing during transformation, *FASEB J.*, 14, 95, 2000.

47. Schubeler, D., Lorincz, M.C., Cimbora, D.M., Telling, A., Feng, Y.Q., Bouhassira, E.E., and Groudine, M., Genomic targeting of methylated DNA: influence of methylation on transcription, replication, chromatin structure, and histone acetylation, *Mol. Cell Biol.*, 20, 9103, 2000.

48. Rice, J.C. and Futscher, B.W., Transcriptional repression of BRCA1 by aberrant cytosine methylation, histone hypoacetylation and chromatin condensation of the BRCA1 promoter. *Nucleic Acids Res.*, 28, 3233,2000.

49. Magdinier, F. and Wolffe, A.P., Selective association of the methyl-CpG binding protein MBD2 with the silent p14/p16 locus in human neoplasia, *Proc. Natl. Acad. Sci. USA*, 98, 4990, 2001.

50. Nguyen, C.T., Gonzales, F.A., and Jones, P.A., Altered chromatin structure associated with methylation-induced gene silencing in cancer cells: correlation of accessibility, methylation, MeCP2 binding and acetylation. *Nucleic Acids Res.*, 29, 4598, 2001.

51. Ballestar, E. et al., Methyl-CpG binding proteins identify novel sites of epigenetic inactivation in human cancer. *EMBO J.*, 22, 6335, 2003.

52. Weinmann, A.S., Bartley, S.M., Zhang, T., Zhang, M.Q., and Farnham, P.J., Use of chromatin immunoprecipitation to clone novel E2F target promoters. *Mol. Cell Biol.,* 21, 6820, 2001.

53. Weinmann, A.S., Yan, P.S., Oberley, M.J., Huang, T.H., and Farnham, P.J., Isolating human transcription factor targets by coupling chromatin immunoprecipitation and CpG island microarray analysis. *Genes Dev.*, 16, 235, 2002.

54. Ren, B. et al., Genome-wide location and function of DNA binding proteins, *Science*, 290, 2306, 2000.

55. Lieb, J.D., Liu, X., Botstein, D., and Brown, P.O., Promoter-specific binding of Rap1 revealed by genome-wide maps of protein-DNA association. *Natl. Genet.*, 28, 327, 2001.

11 Mouse Models for the Study of DNA Methylation

Nicole M. Sodir and Peter W. Laird

CONTENTS

ACKNOWLEDGMENTS

The authors would like to thank Dr. Daniel Weisenberger for his critical reading of the manuscript.

INTRODUCTION

Aberrant DNA methylation is associated with transcriptional silencing of tumor suppressor genes in neoplasia.[1,2] Recently, in addition to DNA methylation, chromatin modifications have been shown to be involved in modulating gene expression.[2] However, the interplay among DNA methyltransferases responsible for DNA methylation, chromatin structure, and chromatin remodeling proteins in neoplasia is not fully understood in mammals. It remains to be determined whether the aberrant signal is initiated first at the chromatin level and then transmitted to the DNA level or vice versa. Many knockout models for the different DNA methyltransferases and chromatin modulators have been developed. This chapter will focus on describing some of these models and their role in enhancing our understanding of DNA methylation and cancer.

DNA METHYLTRANSFERASES

In mammals, DNA methylation occurs at the carbon-5 position of cytosine residues in the context of a CpG dinucleotide pair. There are two distinct types of DNA methyl transfer reactions: *de novo* methylation and maintenance methylation. The process of *de novo* methylation occurs when a CpG site that was previously unmethylated becomes methylated. Maintenance methylation occurs when a methyl group is transferred to a cytosine residue in the daughter strand that has a methylated cytosine at CpG dinucleotide in the parent strand. To date, five mammalian DNA methyltransferases have been identified (Dnmt1, Dnmt2, Dnmt3a, Dnmt3b, and Dnmt3L). Each methyltransferase will be discussed in detail.

DNMT1

Dnmt1 was identified in 1988;[3] however, its sequence remained incomplete for several more years.[4] It is composed of two parts, a diversified amino terminal region and a conserved carboxy terminal region (Figure 11.1). The N-terminal region in Dnmt1 is the largest among all mammalian DNA methyltransferases and has several functional elements, including a nuclear localization signal, replication foci targeting sequences, a zinc binding domain, and a proliferating cell nuclear antigen interacting motif. The C-terminal domain consists of the catalytic domain with all the conserved motifs (I-X) necessary for the activity of the methyltransferase.[5,6] Dnmt1 is commonly referred to as the maintenance methyltransferase because of its low *de novo* activity and its high preference toward hemimethylated DNA. Dnmt1 is required for embryonic development, genomic imprinting, and X-inactivation.[7–9]

DNMT2

Dnmt2 was identified in 1998 and shares homology with the pmt1 protein of the fission yeast *S. pombe*.[10,11] Although it contains all of the conserved methyltransferase motifs, the yeast pmt1 protein is not thought to be a functional methyltransferase since the catalytic Pro-Pro-Cys motif is not conserved, but instead is Pro-Ser-Cys. Restoration of the consensus sequence in pmt1 resulted in a catalytically active methyltransferase.[12] Dnmt2 contains a catalytic domain with all the correct conserved motifs. Nevertheless, human DNMT2 produced in *E.coli* and mouse Dnmt2 produced in a baculovirus system failed to methylate DNA *in vitro*.[10,11] In addition, Dnmt2-deficient embryonic stem (ES) cells retain their ability to *de novo* methylate integrated DNA viruses.[10] Recently, a study by Hermann et al. showed that Dnmt2 has a weak methyltransferase activity observed at CG sites within the ttnCGa(g/a) consensus sequence.[13]

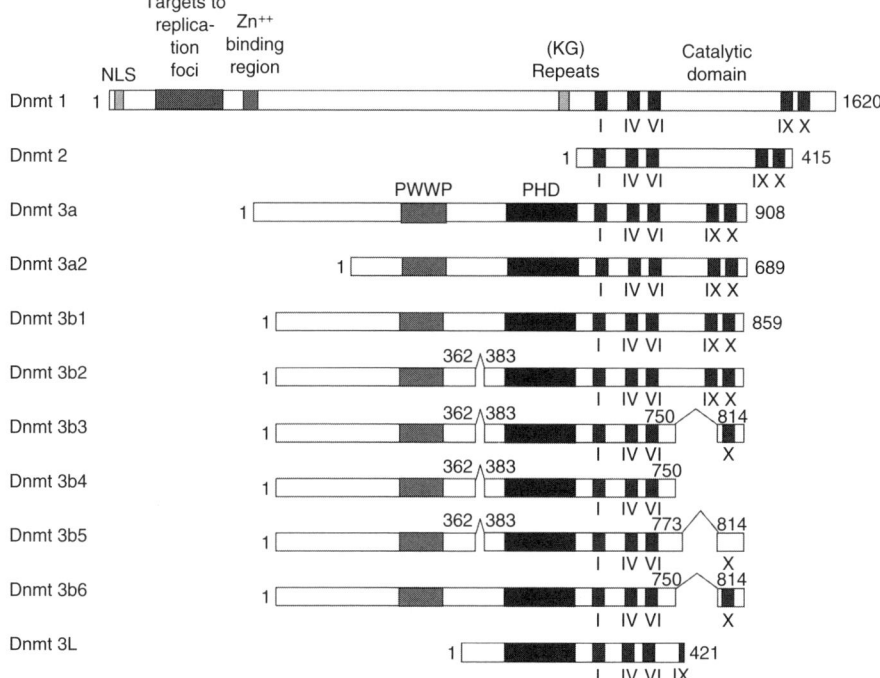

FIGURE 11.1 Schematic diagram of the structure of the murine DNA (cytosine)-methyl-transferase proteins. NLS stands for nuclear localization signal. The replication foci targeting sequences, the Zn^{++}-binding region, the conserved PWWP and PHD motifs, the conserved motifs in the catalytic domain (I, IV, VI, and X), in addition to the sites of alternative splicing, are indicated. (*Source*: Derived in part from Chen, T. et al., *Mol. Cell. Biol.,* 23 (16), 5594–5605, 2003.)

DNMT3A AND DNMT3B

Dnmt 3a and Dnmt3b, identified by Okano et al. in 1998,[14] are highly conserved between human and mouse, but have little sequence homology to either Dnmt1 or Dnmt2. They are expressed in undifferentiated ES cells, but are expressed at a much lower level in differentiated cells and adult tissues.[15] *Dnmt3a* and *Dnmt3b* encode functional proteins that show approximately equal preference for both unmethylated and hemimethylated DNA substrate *in vitro*[14] as well as *in vivo*.[16] Dnmt3a and Dnmt3b are essential for *de novo* methylation in ES cells and during embryogenesis.[17]

The *Dnmt3a* gene encodes two isoforms, Dmnt3a and Dnmt3a2, driven by separate promoters (Figure 11.1). Dnmt3a and Dnmt3a2 proteins share identical amino acid sequences, but they differ at the N-terminal end.[18] Dnmt3a2 is the predominant *Dnmt3a* gene product in ES cells and is mainly localized to euchromatin, whereas Dnmt3a is primarily concentrated in the heterochromatin.[18]

To date, six Dnmt3b isoforms have been identified, which are produced via alternative splicing (Figure 11.1). Dnmt3b1 and Dnmt3b2 are enzymatically active, whereas Dnmt3b3, Dnmt3b4, Dnmt3b5, and Dnmt3b6 are thought to be catalytically inactive due to the lack of the conserved motif IX essential for the methyltransferase activity resulting from alternative splicing between exons 21 and 22.[18,19]

DNMT3L

The mouse *Dnmt3L* gene (DNA methyltransferase 3-like) was identified in 2001 and shares 61% identity with its human counterpart.[20,21] The N-terminal part of Dnmt3L protein has a cysteine-rich region containing a novel-type zinc finger domain, highly conserved in Dnmt3a, Dnmt3b, and ATRX. The C-terminal part of the protein shares similarity with the motifs I, IV, and VI found in all other DNA methyltransferases. However, it lacks the consensus amino acid residues in these motifs needed for methyltransferase activity.[21]

Dnmt3L is expressed in ES cells, in the chorion of E7.5 and E8.5 embryos, in oocytes, and in differentiating spermatocytes of newborn and adult mice.[22] This expression pattern is very similar to that of Dnmt3a and Dnmt3b.[14,17] Dnmt3L has been shown to interact with Dnmt3a and Dnmt3b and to colocalize with these enzymes in the heterochromatin foci of transfected mammalian cells.[22] Dnmt3L is involved in the establishment of maternal genomic imprinting.[22,23] Since Dnmt3L by itself has no intrinsic methyltransferase activity, these data suggest that Dnmt3L regulates genomic imprinting by recruiting Dnmt3a and Dnmt3b to specific loci in the genome.[22] The role of Dnmt3L in DNA methylation has also been assessed in human cells using replicating minichromosomes carrying various imprinting centers (ICs) as methylation targets, in which human DNMT3L was shown to stimulate *de novo* methylation activity of Dnmt3a.[24]

GENE TARGETING OF *Dnmt* METHYLTRANSFERASES

Many targeted mutations of the murine *Dnmt1* gene have been generated, which differ in the location of the targeted region. The resulting alleles differ qualitatively and quantitatively in their residual Dnmt1 expression. The *Dnmt1ⁿ* allele (N-terminal disruption or *NaeI* deletion), the most studied *Dnmt1* hypomorphic allele, has a neomycin expression cassette that replaces part of exon 4 at *NaeI* site. *Dnmt1ⁿ/ⁿ* embryos are severely stunted, developmentally delayed, and die at E10.5. They also show severe DNA hypomethylation with 30% reduction in the normal content of methylcytosines.[7,9]

The *Dnmtˡˢ* allele (*SalI* site insertion) is a null allele with a neomycin expression cassette inserted at *SalI* site in the replication foci-targeting region in the N-terminal domain.[9] *Dnmt1ᶜ* (catalytic domain disruption) is a null allele generated by deletion of two conserved motifs in the catalytic domain. Mice heterozygous for either *Dnmt1ˢ* or *Dnmt1ᶜ* are viable and fertile; however, both *Dnmt1ˢ/ˢ* and *Dnmt1ᶜ/ᶜ* embryos show significantly lower methylation levels than the *Dnmt1ⁿ/ⁿ* embryos and die at E8.5.[9] ES cells homozygous for *Dnmt1ⁿ*, *Dnmt1ˢ*, or *Dnmt1ᶜ* alleles with extensive DNA hypomethylation are viable and grow normally in the undifferentiated state; however, they die upon differentiation.[9]

Another *Dnmt1* targeting event, *Dnmt1^Chip^*, was generated by insertion of a genomic/cDNA *Dnmt1* hybrid into the endogenous locus of *Dnmt1*.[25] This results in functional Dnmt1 transcripts but at levels lower than those of the wild type *Dnmt1* allele. *Dnmt1^Chip^* has been used in combination with severe knockout alleles like *Dnmt1^c^*, which will be discussed later.[26,27]

The *Dnmt1^r^* allele (EcoRV insertion) contains a 320-bp insertion containing three copies of the lac operator sequence (*lacO*) from *E.coli* located in intron 3 just upstream of the splice acceptor site. *Dnmt1^r^* is a hypomorphic allele that is expressed at a lower level than the wild type allele but at a higher level than the *Dnmt1^n^* allele. *Dnmt1^r/r^* mice are viable and undergo normal development.[28,29]

*Dnmt1^2lox^* was generated through conditional inactivation of *Dnmt1* using the Cre-*loxP* system and has *loxP* sites flanking exons 4 and 5. Cre-mediated deletion of *Dnmt1^2lox^* in primary mouse embryonic fibroblasts resulted in extensive genomic hypomethylation and p53-dependent cell death within 6 days of *Dnmt1* deletion.[30] This cre/lox binary system was also used *in vivo* to study the role of Dnmt1 in the development of the nervous system. Conditional knockout mice in which *Dnmt1* deletion achieved in either postmitotic CNS neurons or in E9-E12 CNS precursor cells were generated. Dnmt1 deficiency had no effect on postmitotic neurons. On the other hand, Dnmt1 deficiency in mitotic precursor cells resulted in significant DNA hypomethylation of progeny cells. Mutant mice with 95% hypomethylated cells in the brain died postnatally of respiratory distress, whereas mosaic mice with 30% hypomethylated CNS cells survived until adulthood.[31]

Null mutations of *Dnmt3a* and *Dnmt3b* genes have also been generated by deleting part of the catalytic domain containing two highly conserved motifs.[17] *Dnmt3a^-/-^* mice develop to term; however, although they appear to be normal at birth, most of them were stunted and died at about 4 weeks of age. *Dnmt3b^-/-^* mice are not viable as the embryos suffer from multiple developmental defects, including growth impairment between 9.5 and 16.5 dpc.[17]

The *Dnmt2* gene has been inactivated in ES cells by targeted deletion of a conserved motif in the catalytic domain. Dnmt2-deficient ES cells displayed normal growth and morphology with no significant change in their methylation status.[10]

Targeted disruption of *Dnmt3L* was performed independently by two groups.[22,23] *Dnm3L^-/-^* mice are viable and survive to adulthood without any visible morphological abnormalities. However, both sexes are infertile. *Dnmt3L^-/-^* males have azoospermia, while the heterozygous progeny of *Dnmt3L^-/-^* females have developmental defects and die before midgestation.[22,23]

DNA METHYLATION AND CHROMATIN REMODELING

Gene silencing in neoplasia is associated with alterations of both DNA methylation patterns and chromatin structure. However, it is still unclear whether methylation initiates this process or whether it is a result of chromatin remodeling events. Evidence for the linkage between DNA methylation and chromatin configuration will be discussed below in *Lsh* and *Mbd2* knockout mouse models.

Lsh Knockout Mice

Lsh (lymphoid-specific helicase) is a member of the SNF2 chromatin remodeling family. The SNF2 family of proteins utilizes the energy derived from ATP hydrolysis and disrupts histone/DNA interactions allowing the sliding of nucleosomes on the DNA.[32,33] *Lsh* is highly expressed in lymphoid tissues in adult mice and has been shown to be essential for their normal development.[34,35] *Lsh* shares 50% identity in the helicase region with Ddm1 (decrease in DNA methylation 1), another member of SNF2 family. Ddm1 acts as a modulator of genomic methylation in *Arabidopsis thaliana*; its mutation results in 70% loss of the 5-methylcytosine content.[36,37] *Lsh*-deficient mice develop to term but die shortly after birth. They have abnormal lymphoid development, renal defects, and a 20% reduction of birth weight. They also show defects in CpG methylation throughout the genome.[38] Analysis of repetitive elements, including minor and major satellites sequences, Sine B1 repeats and telomeric sequences, as well as single copy loci like β-*Globin* and *Pgk-1* genes, all of which are heavily methylated in normal cells, revealed substantial hypomethylation in all the examined tissues in the absence of *Lsh*.

Lsh is a nuclear protein that is shown to localize to the pericentric heterochromatin.[39] Under normal circumstances, heterochromatin composed of hypermethylated DNA has low methylation at H3-K4 (histone3-lysine4) and high methylation at H3-K9 (histone3-lysine9).[40] The loss of *Lsh* results in the abnormal accumulation of di- and trimethylated H3-K4 at pericentromeric DNA and other repetitive sequences, and in the reactivation of the retroviral gene expression.[41] In addition, disruption of higher order heterochromatin organization by treatment with the histone deacetylase inhibitor Trichostatin A (TSA) leads to the dissociation of *Lsh* from pericentromeric heterochromatin,[39] suggesting that *Lsh* might be recruited to an intact heterochromatin region in order to maintain normal heterochromatin formation. As a result, *Lsh* is crucial for DNA and histone methylation, and *Lsh*-deficient mice provide an important model system to characterize the relationship between DNA methylation and chromatin structure.

Mbd2 Knockout Mice

Mbd2 (methyl-CpG binding repressor 2) binds to methylated DNA and represses transcription through the recruitment of corepressor complexes containing histone deacetylases.[42] *Mbd2*[-/-] mice are viable, fertile, and contain normal global levels of genomic CpG methylation.[43] However, they are defective in the silencing of certain developmentally regulated genes[44] and have abnormal maternal behavior; *Mbd2*[-/-] mothers fail to adequately feed their pups.[43]

Silencing of tumor suppressor genes through CpG methylation at their promoter regions contributes to cancer.[1] Since Mbd2 interprets the DNA methylation signal through transcriptional repression, the effect of Mbd2 deficiency on tumorigenesis was tested on the Min (multiple intestinal neoplasia) mouse model system.[45] Min mice carry a germline mutation in the *Apc* tumor suppressor gene and develop multiple intestinal adenomas in the first 6 months of life.[46,47] The deficiency of Mbd2 increased *Apc*[Min/+] mouse survival and decreased tumor burden in a dose-dependent manner. $Apc^{Min/+}$ $Mbd2^{-/-}$ mice live significantly longer (median = 354 days) than $Apc^{Min/+}$ $Mbd2^{+/-}$ mice (median = 238 days) or $Apc^{Min/+}$ $Mbd2^{+/+}$ mice (median = 183 days).[45] In addition, $Apc^{Min/+}$ $Mbd2^{-/-}$ mice had approximately 10 times fewer adenomas than $Apc^{Min/+}$ $Mbd2^{+/-}$ control mice. Another effect of Mbd2 deficiency is the altered distribution of intestinal tumors. The few adenomas that arise in $Apc^{Min/+}$ $Mbd2^{-/-}$ mice are mainly located in the distal colon compared to the small intestine in $Apc^{Min/+}$ $Mbd2^{+/+}$ mice.[45] This study showed that Mbd2 is involved in the Min tumorigenic pathway and emphasized the direct contribution of both aberrant DNA methylation and chromatin remodeling events in the process of tumorigenesis.

HYPOMETHYLATION AND CANCER

DNA methylation alterations are common in human tumors.[2] At the global level, the genomic content of 5-methylcytosine is often decreased in tumor cells compared to normal cells;[48] however, DNA hypermethylation at the promoters of tumor suppressor genes is common and is associated with transcriptional repression and thus neoplasia.[1,2] The function of cancer-linked DNA hypomethylation is not well understood.

The biological and the functional significance of DNA hypomethylation in cancer were explored by *in vivo* modulation of DNA methylation, accomplished through genetic manipulation of the *Dnmt1* gene. Since homozygous knockout *Dnmt1* mice die during gestation, Laird et al. used the combined effect of heterozygosity for a null mutation of the *Dnmt1* gene ($Dnmt1^S$), and the treatment with low doses of the DNA methyltransferase inhibitor 5-aza-2′-deoxycytidine (5-azaCdR) to reduce the methyltransferase activity in the Min mouse model system.[49] The decreased methylation in Min mice led to a significant reduction in the average number of intestinal adenomas. Control $Dnmt1^{+/+}$ $Apc^{Min/+}$ mice developed an average of 113 intestinal adenomas, while heterozygous $Dnmt1^{S/+}$ $Apc^{Min/+}$ mice treated with 5-azaCdR and thus with substantial hypomethylation in their tissues developed an average of 2 intestinal adenomas. One caveat to this study is that the reduction in intestinal adenomas could be caused in part by 5-azaCdR cytotoxicity and not by DNA hypomethylation.[50] To circumvent this issue, Eads et al. used a pure genetic approach by introducing two hypomorphic alleles, $Dnmt1^N$ and $Dnmt1^R$, into the Min mouse background.[51] $Dnmt^{N/R}$ $Apc^{Min/+}$ mice with substantially reduced DNA methylation were completely protected from polyp development, demonstrating the complete dependence of Min polyp formation on sufficient levels of DNA methyltransferase expression. To address whether this is a broad phenomenon, Trinh et al. used a similar strategy to introduce $Dnmt1^N$ and $Dnmt1^R$ hypomorphic alleles into a different mouse model that does not involve Apc^{Min}, the Mlh1 mismatch repair deficient mice.[28] $Mlh1^{-/-}$ mice are predisposed to the development of multiple tumors, including lymphomas and intestinal adenocarcinomas.[52] *Dnmt1* hypomorphic $Mlh1^{-/-}$ mice showed a diminished frequency of intestinal tumors and an exacerbated frequency of aggressive lymphoid tumors. This suggests that the effect of hypomethylation on tumor development is dependent on the tissue of origin that gives rise to the tumor. In a recent study, Gaudet et al. generated mice with a 90% reduced level of Dnmt1 expression by combining the hypomorphic allele $Dnmt1^{chip}$ with the null allele $Dnmt^C$.[26] The $Dnmt1^{chip/C}$ compound heterozygote mice had substantial genome-wide hypomethylation in all tissues and at 4 to 8 months of age developed aggressive T-cell lymphomas that exhibited a high frequency of chromosome 15 trisomy. This suggests that DNA hypomethylation induces tumor formation by promoting chromosomal instability.

DNA hypomethylation has been linked to chromosomal instability in multiple studies. Patients with ICF (immunodeficiency, centromere instability, and facial anomalies) syndrome, with mutations in the catalytic domain of Dnmt3b, show profound chromosomal abnormalities including decondensation of the centromeric and pericentromeric regions of particular chromosomes, associated with methylation-deficient repetitive sequences in the region.[53,54] In studies involving colorectal cancer cell lines, a striking correlation between genetic instability and methylation capacity was observed, indicating that defects in DNA methylation might contribute to abnormal chromosomal segregation in cancer cells.[55] In addition, it was shown that the loss of the murine Su(var)3-9 methyltransfersase, which governs H3-K9 methylation at pericentric heterochromatin, results directly in chromosomal instabilities that are associated with increased risk of B-cell lymphomas and perturbed chromosomal segregation during male meiosis.[56]

The effect of hypomethylation on genomic integrity was also evaluated in murine ES cells in two separate studies. Chen et al. first showed that homozygous $Dnmt1$ knockout ES cells with global DNA hypomethylation have increased aberrant recombination events and, therefore, elevated mutation rates at the endogenous X-linked hypoxanthine phosphoribosyltransferase ($Hprt$) and integrated autosomal viral thymidine kinase (Tk) genes.[57] However, Chan et al. used random integrations of a chimeric $TKNeo$ (thymidine kinase neomycin phosphotransferase) transgene to select for mutations and found that Dnmt1-deficicient ES cells have lower gene inactivation and mutation rates at the $TKNeo$ loci than Dnmt1-proficient cells.[58] This discrepancy might be caused in part by differences in the chromosomal locations assessed in the two studies.

To further explore the effect of hypomethylation on genomic instability, Eden et al. used tumor-prone mice harboring mutations in the neurofibromatosis $(Nf1)$ and $p53$ alleles linked on mouse chromosome 11.[27] All mice carrying mutations of $Nf1$ and $p53$ alleles in cis ($Npcis$) develop soft tissue sarcomas between 3 and 7 months of age that exhibit loss of heterozygosity (LOH) at both gene loci.[59] The advantage of this model system is LOH of both alleles can be easily scored in a tissue culture system. $Npcis$ mice carrying the hypomorphic allele $Dnmt1^{Chip/-}$ develop sarcomas at an earlier age and have a roughly twofold increase in LOH events compared with $Npcis$ $Dnmt1^{+/+}$ mice. This study provides the most direct evidence of the role of DNA hypomethylation in promoting gene inactivation and genomic instability by LOH. However, the mechanisms by which hypomethylation promotes LOH are still unknown. Further studies will need to address the relationship between hypomethylation and other mechanisms of gene inactivation, like gene mutation and gene silencing, in somatic cells that are more relevant to the process of carcinogenesis than ES cells.

The reduction of DNA methylation seems to have opposing effects in the development of different types of cancer. Hypomethylation can simultaneously enhance soft tissue sarcomas and the rate of lymphomagenesis and reduce the incidence of intestinal adenomas in Min mice. Such discrepancies in the outcomes of hypomethylation on tumor formation may be due to varying mechanisms or pathways in the tissues of origin that gave rise to the tumors. A better understanding of these pathways is necessary to choose the appropriate therapeutic strategy for cancer patients.

OUTLOOK

The embryonic lethality of *Dnmt1* knockout mice has prevented the analysis of Dnmt1 function in adult tissues and in neoplasia. To overcome this problem, *Dnmt1* heterozygous and hypomorphic mice have been developed, which has facilitated understanding of the role of DNA methylation in various mouse models of cancer. A drawback of these systems is that the modulation of DNA methylation in the germline may affect multiple tissues simultaneously, precluding the detailed study of its function on a specific cell lineage in isolation. In addition, the study of the effect of DNA methylation on a tissue-specific type of cancer might be complicated by the severity of phenotype in another tissue. For instance, $Dnmt1^{N/R} Mlh1^{-/-}$ mice had an exacerbated lymphomagenesis that caused earlier lethality and made difficult the study of intestinal cancer.[28]

The next step will be to circumvent these limitations by generating conditional knockout mice in which the loss of function of Dnmt1 is induced in a tissue-restricted and time-specific manner. The Cre-*loxP* system has been successfully used to produce tissue-restricted expression of Dnmt1.[30] However, this system is of limited use because cre-based recombination is a highly stochastic event in which a subset of cells undergoes complete inactivation of Dnmt1. Differentiated cells that have lost Dnmt1 expression undergo p53-dependent cell death,[30] which makes it difficult to study the early stages of neoplasia *in vivo*. As an alternative, our laboratory is using another strategy of developing an *in vivo* tissue-specific transcriptional repression of *Dnmt1* based on a prokaryotic repressor/operator system. This system allows most of the cells to undergo partial inactivation of Dnmt1.

In addition to *Dnmt1* conditional knockout mice, the generation of conditional knockouts of other DNA methyltransferases (Dnmt2, Dnmt3a, Dnmt3b, and Dnmt3L) and chromatin remodeling proteins (*Lsh* and Mbd2) will be useful in elucidating the mechanisms by which DNA methylation affects tumorigenesis. Moreover, the generation of transgenic mice with increased expression of these genes in a conditional manner, and the usage of mouse imaging systems to noninvasively monitor tumor development in these mouse models, will complement our understanding of these mechanisms.

REFERENCES

1. Jones, P. A. and Laird, P. W., Cancer epigenetics comes of age, *Nat. Genet.,* 21, 163–167, 1999.
2. Jones, P. A. and Baylin, S. B., The fundamental role of epigenetic events in cancer, *Nat. Rev. Genet.,* 3, 415–428, 2002.

3. Bestor, T., Laudano, A., Mattaliano, R., and Ingram, V., Cloning and sequencing of a cDNA encoding DNA methyltransferase of mouse cells. The carboxyl-terminal domain of the mammalian enzymes is related to bacterial restriction methyltransferases, *J. Mol. Biol.*, 203, 971–983, 1988.

4. Yoder, J. A., Yen, R. W., Vertino, P. M., Bestor, T. H., and Baylin, S. B., New 5′ regions of the murine and human genes for DNA (cytosine-5)-methyltransferase, *J. Biol. Chem.*, 271, 31092–31097, 1996.

5. Bestor, T. H., The DNA methyltransferases of mammals, *Hum. Mol. Genet.*, 9, 2395–2402, 2000.

6. Robertson, K. D., DNA methylation, methyltransferases, and cancer, *Oncogene*, 20, 3139–3155, 2001.

7. Li, E., Bestor, T. H., and Jaenisch, R., Targeted mutation of the DNA methyltransferase gene results in embryonic lethality, *Cell*, 69, 915–926, 1992.

8. Beard, C., Li, E., and Jaenisch, R., Loss of methylation activates Xist in somatic but not in embryonic cells, *Genes Dev.*, 9, 2325–2334, 1995.

9. Lei, H., Oh, S. P., Okano, M., Juttermann, R., Goss, K. A., Jaenisch, R., and Li, E., De novo DNA cytosine methyltransferase activities in mouse embryonic stem cells, *Development*, 122, 3195–3205, 1996.

10. Okano, M., Xie, S., and Li, E., Dnmt2 is not required for de novo and maintenance methylation of viral DNA in embryonic stem cells, *Nucleic Acids Res.*, 26, 2536–2540, 1998.

11. Yoder, J. A. and Bestor, T. H., A candidate mammalian DNA methyltransferase related to pmt1p of fission yeast, *Hum. Mol. Genet.*, 7, 279–284, 1998.

12. Pinarbasi, E., Elliott, J., and Hornby, D. P., Activation of a yeast pseudo DNA methyltransferase by deletion of a single amino acid, *J. Mol. Biol.*, 257, 804–813, 1996.

13. Hermann, A., Schmitt, S., and Jeltsch, A., The human Dnmt2 has residual DNA-(cytosine-C5) methyltransferase activity, *J. Biol. Chem.*, 278 (34), 31717–31721, 2003.

14. Okano, M., Xie, S., and Li, E., Cloning and characterization of a family of novel mammalian DNA (cytosine-5) methyltransferases, *Nat. Genet.*, 19, 219–220, 1998.

15. Robertson, K. D., Uzvolgyi, E., Liang, G., Talmadge, C., Sumegi, J., Gonzales, F. A., and Jones, P. A., The human DNA methyltransferases (DNMTs) 1, 3a and 3b: coordinate mRNA expression in normal tissues and overexpression in tumors, *Nucleic Acids Res.*, 27, 2291–2298, 1999.

16. Hsieh, C. L., In vivo activity of murine de novo methyltransferases, Dnmt3a and Dnmt3b, *Mol. Cell. Biol.*, 19 (12), 8211–8218, 1999.

17. Okano, M., Bell, D. W., Haber, D. A., and Li, E., DNA methyltransferases Dnmt3a and Dnmt3b are essential for de novo methylation and mammalian development, *Cell*, 99 (3), 247–257, 1999.

18. Chen, T., Ueda, Y., Xie, S., and Li, E., A novel Dnmt3a isoform produced from an alternative promoter localizes to euchromatin and its expression correlates with active de novo methylation, *J. Biol. Chem.*, 277, 38746–38754, 2002.

19. Chen, T., Ueda, Y., Dodge, J. E., Wang, Z., and Li, E., Establishment and maintenance of genomic methylation patterns in mouse embryonic stem cells by Dnmt3a and Dnmt3b, *Mol. Cell. Biol.*, 23 (16), 5594–5605, 2003.

20. Aapola, U., Kawasaki, K., Scott, H. S., Ollila, J., Vihinen, M., Heino, M., Shintani, A., Minoshima, S., Krohn, K., Antonarakis, S. E., Shimizu, N., Kudoh, J., and Peterson, P., Isolation and initial characterization of a novel zinc finger gene, DNMT3L, on 21q22.3, related to the cytosine-5-methyltransferase 3 gene family, *Genomics*, 65, 293–298, 2000.

21. Aapola, U., Lyle, R., Krohn, K., Antonarakis, S. E., and Peterson, P., Isolation and initial characterization of the mouse Dnmt3l gene, *Cytogenet. Cell Genet.*, 92, 122–126, 2001.

22. Hata, K., Okano, M., Lei, H., and Li, E., Dnmt3L cooperates with the Dnmt3 family of de novo DNA methyltransferases to establish maternal imprints in mice, *Development*, 129 (8), 1983–1993, 2002.

23. Bourc'his, D., Xu, G. L., Lin, C. S., Bollman, B., and Bestor, T. H., Dnmt3L and the establishment of maternal genomic imprints, *Science*, 294, 2536–2539, 2001.

24. Chedin, F., Lieber, M. R., and Hsieh, C. L., The DNA methyltransferase-like protein DNMT3L stimulates de novo methylation by Dnmt3a, *Proc. Natl. Acad. Sci. USA*, 99, 16916–16921, 2002.

25. Tucker, K. L., Beard, C., Dausmann, J., Jackson-Grusby, L., Laird, P. W., Lei, H., Li, E., and Jaenisch, R., Germ-line passage is required for establishment of methylation and expression patterns of imprinted but not of nonimprinted genes, *Genes Dev.*, 10, 1008–1020, 1996.

26. Gaudet, F., Hodgson, J. G., Eden, A., Jackson-Grusby, L., Dausman, J., Gray, J. W., Leonhardt, H., and Jaenisch, R., Induction of tumors in mice by genomic hypomethylation, *Science*, 300 (5618), 489–492, 2003.

27. Eden, A., Gaudet, F., Waghmare, A., and Jaenisch, R., Chromosomal instability and tumors promoted by DNA hypomethylation, *Science*, 300 (5618), 455, 2003.

28. Trinh, B. N., Long, T. I., Nickel, A. E., Shibata, D., and Laird, P. W., DNA methyltransferase deficiency modifies cancer susceptibility in mice lacking DNA mismatch repair, *Mol. Cell. Biol.*, 22, 2906–2917, 2002.

29. Eads, C. A., Lord, R. V., Wickramasinghe, K., Long, T. I., Kurumboor, S. K., Bernstein, L., Peters, J. H., DeMeester, S. R., DeMeester, T. R., Skinner, K. A., and Laird, P. W., Epigenetic patterns in the progression of esophageal adenocarcinoma, *Cancer Res.*, 61, 3410–3418, 2001.

30. Jackson-Grusby, L., Beard, C., Possemato, R., Tudor, M., Fambrough, D., Csankovszki, G., Dausman, J., Lee, P., Wilson, C., Lander, E., and Jaenisch, R., Loss of genomic methylation causes p53-dependent apoptosis and epigenetic deregulation, *Nat. Genet.*, 27, 31–39, 2001.

31. Fan, G., Beard, C., Chen, R. Z., Csankovszki, G., Sun, Y., Siniaia, M., Biniszkiewicz, D., Bates, B., Lee, P. P., Kuhn, R., Trumpp, A., Poon, C., Wilson, C. B., and Jaenisch, R., DNA hypomethylation perturbs the function and survival of CNS neurons in postnatal animals, *J. Neurosci.*, 21, 788–797, 2001.

32. Peterson, C. L., Chromatin remodeling enzymes: taming the machines. Third in review series on chromatin dynamics, *EMBO Rep.*, 3 (4), 319–322, 2002.

33. Becker, P. B. and Horz, W., ATP-dependent nucleosome remodeling, *Annu. Rev. Biochem.*, 71, 247–273, 2002.

34. Geiman, T. M., Durum, S. K., and Muegge, K., Characterization of gene expression, genomic structure, and chromosomal localization of Hells (Lsh), *Genomics*, 54 (3), 477–483, 1998.

35. Geiman, T. M. and Muegge, K., Lsh, An SNF2/helicase family member, is required for proliferation of mature T lymphocytes, *Proc. Natl. Acad. Sci. USA*, 97 (9), 4772–4777, 2000.

36. Jeddeloh, J. A., Stokes, T. L., and Richards, E. J., Maintenance of genomic methylation requires a SWI2/SNF2-like protein, *Nat. Genet.*, 22 (1), 94–97, 1999.

37. Martienssen, R. and Henikoff, S., The House & Garden guide to chromatin remodelling, *Nat. Genet.*, 22 (1), 6–7, 1999.

38. Dennis, K., Fan, T., Geiman, T., Yan, Q., and Muegge, K., Lsh, A member of the SNF2 family is required for genome-wide methylation, *Genes Dev.,* 15, 2940–2944, 2001.

39. Yan, Q., Cho, E., Lockett, S., and Muegge, K., Association of Lsh, a regulator of DNA methylation, with pericentromeric heterochromatin is dependent on intact heterochromatin, *Mol. Cell. Biol.,* 23 (23), 8416–8428, 2003.

40. Lachner, M. and Jenuwein, T., The many faces of histone lysine methylation, *Curr. Opin. Cell Biol.,* 14, 286–298, 2002.

41. Yan, Q., Huang, J., Fan, T., Zhu, H., and Muegge, K., Lsh, a modulator of CpG methylation, is crucial for normal histone methylation, *Embo J.,* 22 (19), 5154–5162, 2003.

42. Ng, H. H., Zhang, Y., Hendrich, B., Johnson, C. A., Turner, B. M., Erdjument-Bromage, H., Tempst, P., Reinberg, D., and Bird, A., MBD2 is a transcriptional repressor belonging to the MeCP1 histone deacetylase complex, *Nat. Genet,* 23 (1), 58–61, 1999.

43. Hendrich, B., Guy, J., Ramsahoye, B., Wilson, V. A., and Bird, A., Closely related proteins MBD2 and MBD3 play distinctive but interacting roles in mouse development, *Genes Dev.,* 15, 710–723, 2001.

44. Hutchins, A. S., Mullen, A. C., Lee, H. W., Sykes, K. J., High, F. A., Hendrich, B. D., Bird, A. P., and Reiner, S. L., Gene silencing quantitatively controls the function of a developmental trans-activator, *Mol. Cell,* 10, 81–91, 2002.

45. Sansom, O. J., Berger, J., Bishop, S. M., Hendrich, B., Bird, A., and Clarke, A. R., Deficiency of Mbd2 suppresses intestinal tumorigenesis, *Nat. Genet.,* 34 (2), 145–147, 2003.

46. Moser, A. R., Dove, W. F., Roth, K. A., and Gordon, J. I., The Min (multiple intestinal neoplasia) mutation: its effect on gut epithelial cell differentiation and interaction with a modifier system, *J. Cell Biol.,* 116, 1517–1526, 1992.

47. Su, L. K., Kinzler, K. W., Vogelstein, B., Preisinger, A. C., Moser, A. R., Luongo, C., Gould, K. A., and Dove, W. F., Multiple intestinal neoplasia caused by a mutation in the murine homolog of the APC gene [published erratum appears in *Science* 1992 May 22; 256(5060):1114], *Science,* 256, 668–670, 1992.

48. Ehrlich, M., DNA methylation in cancer: too much, but also too little, *Oncogene,* 21, 5400–5413, 2002.

49. Laird, P. W., Jackson-Grusby, L., Fazeli, A., Dickinson, S. L., Jung, W. E., Li, E., Weinberg, R. A., and Jaenisch, R., Suppression of intestinal neoplasia by DNA hypomethylation, *Cell,* 81, 197–205, 1995.

50. Juttermann, R., Li, E., and Jaenisch, R., Toxicity of 5-aza-2'-deoxycytidine to mammalian cells is mediated primarily by covalent trapping of DNA methyltransferase rather than DNA demethylation, *Proc. Natl. Acad. Sci. USA,* 91, 11797–11801, 1994.

51. Eads, C. A., Nickel, A. E., and Laird, P. W., Complete genetic suppression of polyp formation and reduction of CpG-island hypermethylation in Apc(Min/+) Dnmt1-hypomorphic Mice, *Cancer Res.,* 62, 1296–1299, 2002.

52. Baker, S. M., Plug, A. W., Prolla, T. A., Bronner, C. E., Harris, A. C., Yao, X., Christie, D. M., Monell, C., Arnheim, N., Bradley, A., Ashley, T., and Liskay, R. M., Involvement of mouse Mlh1 in DNA mismatch repair and meiotic crossing over, *Nat. Genet.,* 13, 336–342, 1996.

53. Jeanpierre, M., Turleau, C., Aurias, A., Prieur, M., Ledeist, F., Fischer, A., and Viegas-Pequignot, E., An embryonic-like methylation pattern of classical satellite DNA is observed in ICF syndrome, *Hum. Mol. Genet.,* 2, 731–735, 1993.

54. Xu, G. L., Bestor, T. H., Bourc'his, D., Hsieh, C. L., Tommerup, N., Bugge, M., Hulten, M., Qu, X., Russo, J. J., and Viegas-Pequignot, E., Chromosome instability and immunodeficiency syndrome caused by mutations in a DNA methyltransferase gene, *Nature,* 402 (6758), 187–191, 1999.

55. Lengauer, C., Kinzler, K. W., and Vogelstein, B., DNA methylation and genetic instability in colorectal cancer cells, *Proc. Natl. Acad. Sci. USA,* 94, 2545–2550, 1997.

56. Peters, A. H., O'Carroll, D., Scherthan, H., Mechtler, K., Sauer, S., Schofer, C., Weipoltshammer, K., Pagani, M., Lachner, M., Kohlmaier, A., Opravil, S., Doyle, M., Sibilia, M., and Jenuwein, T., Loss of the Suv39h histone methyltransferases impairs mammalian heterochromatin and genome stability, *Cell,* 107 (3), 323–337, 2001.

57. Chen, R. Z., Pettersson, U., Beard, C., Jackson-Grusby, L., and Jaenisch, R., DNA hypomethylation leads to elevated mutation rates, *Nature,* 395, 89–93, 1998.

58. Chan, M. F., van Amerongen, R., Nijjar, T., Cuppen, E., Jones, P. A., and Laird, P. W., Reduced rates of gene loss, gene silencing, and gene mutation in Dnmt1-deficient embryonic stem cells, *Mol. Cell. Biol.,* 21, 7587–7600, 2001.

59. Vogel, K. S., Klesse, L. J., Velasco-Miguel, S., Meyers, K., Rushing, E. J., and Parada, L. F., Mouse tumor model for neurofibromatosis type 1, *Science,* 286 (5447), 2176–2179, 1999.

12 DNA Demethylating Agents: Concepts

Jonathan C. Cheng, Daniel J. Weisenberger, and Peter A. Jones

CONTENTS

INTRODUCTION

During tumorigenesis, abnormal cytosine methylation within CpG islands has been demonstrated to be strongly associated with the transcriptional repression of critical genes involved in various cellular processes. Genes involved in all facets of tumor development and progression can become abnormally methylated and epigenetically silenced. Since cytosine methylation is a biochemical epigenetic modification of DNA that does not result in changes in the genetic code, the inactivation of gene expression by DNA methylation is, thus, a process that can be modulated by biochemical or biological manipulations. Given the ubiquitous nature and reversibility of this phenomenon, it is appealing to design strategies to reverse and reestablish the various important functions of tumor suppressor or cancer-related genes that

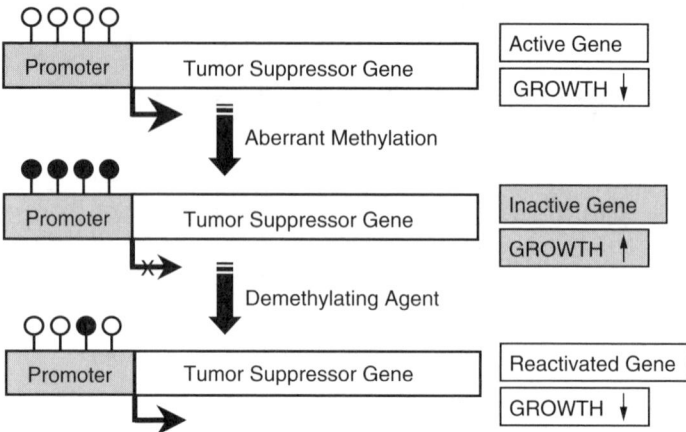

FIGURE 12.1 Model for inactivation of a tumor suppressor gene by aberrant methylation of its promoter region. Normally, active tumor suppressor genes are unmethylated in their promoter/5' regions, hence functioning to suppress cell growth. During tumorigenesis, however, many tumor suppressor genes become abnormally methylated in their promoter/5' regions, thereby leading to the inactivation and silencing of these genes. Nevertheless, demethylating agents can be utilized to demethylate the promoter/5' regions of these genes, leading to reactivation of genes and reconstitution of cell growth control.

have been silenced by methylation. Reversal of expression of epigenetically silenced genes may result in tumor growth suppression and increased sensitivity to anticancer drugs, thus making the application of pharmacologic inhibitors of DNA methylation a conceptually attractive and rational approach for anticancer treatment (Figure 12.1). Agents that can reverse DNA methylation include nucleoside and non-nucleoside inhibitors of DNA methylation. In this chapter, we will discuss various demethylating agents and the concepts behind their application.

INHIBITORS OF DNA METHYLATION

NUCLEOSIDE ANALOGS

The most well-known inhibitors of DNA methylation include the nucleoside analogs, 5-azacytidine (5-Aza-CR), which was originally synthesized as a cancer chemotherapeutic agent in 1964 (Sorm et al., 1964), and its deoxy analog, 5-aza-2-deoxycytidine (5-Aza-CdR). These compounds are ring analogs of the pyrimidine nucleosides, cytidine and 2-deoxycytidine, respectively, but differ by having nitrogen atoms in place of carbons in the fifth position of the ring (Figure 12.2) (Sorm et al., 1964; Vesely, Cihak and Sorm, 1968). The primary metabolic activation of 5-Aza-CR is mediated by uridine–cytidine kinase, whereas 5-Aza-CdR is activated by deoxycytidine kinase. Following their phosphorylation to monophosphates, both compounds are eventually incorporated into newly synthesized DNA. Once incorporated into the DNA, both compounds can form a covalent complex with the major DNA methyltransferase (DNMT1) and thereby trap the enzyme (Figure 12.3A)

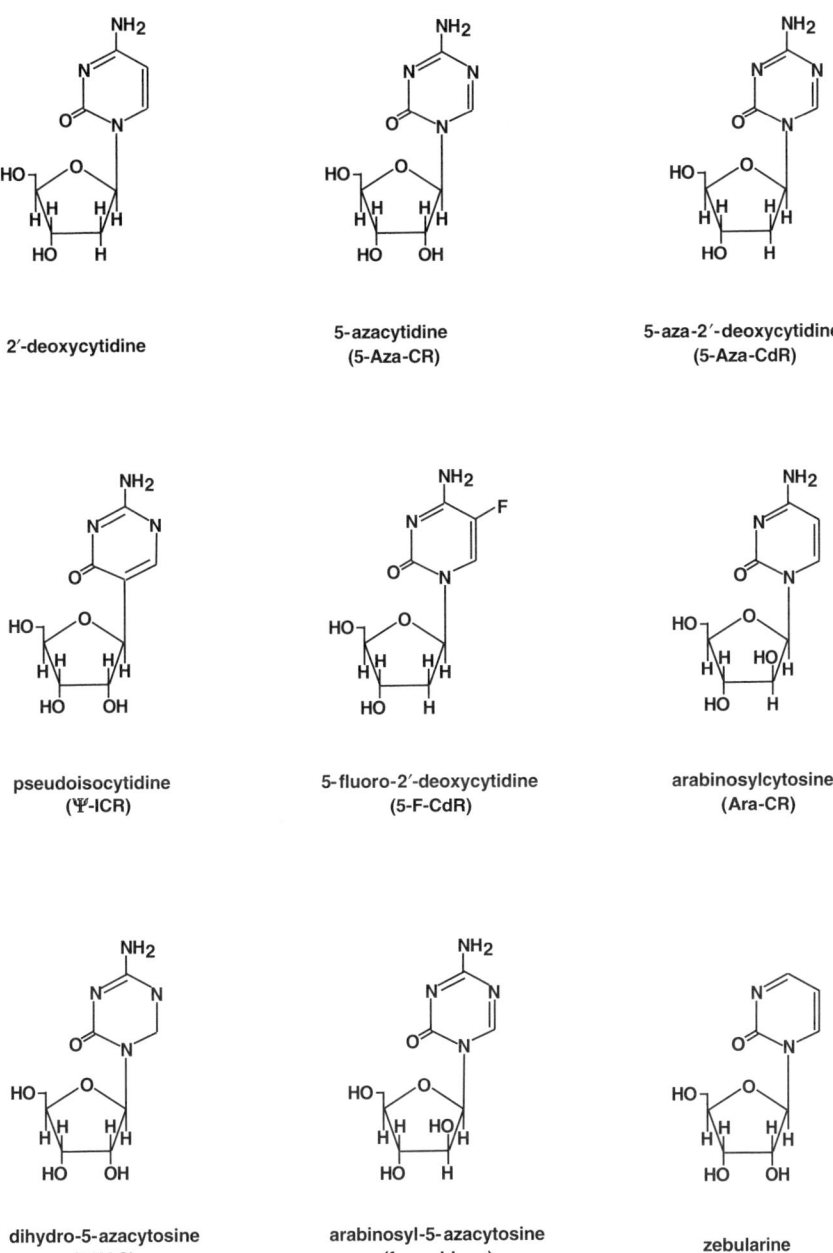

FIGURE 12.2 Chemical structures of cytidine and deoxycytidine analogs. Of the eight cytidine and deoxycytidine analogs, only arabinosylcytidine (Ara-CR) was not shown to be an inhibitor of DNA methylation.

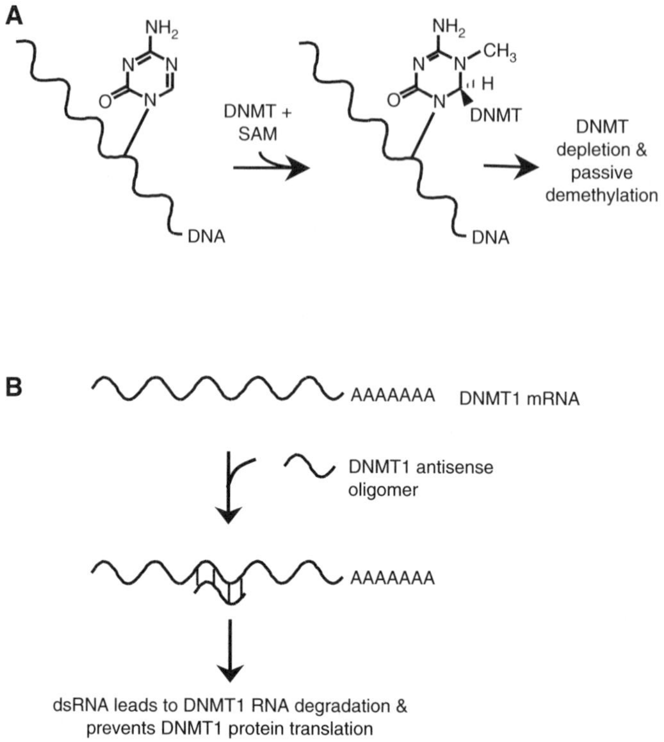

FIGURE 12.3 Mechanisms of DNA methyltransferase (DNMT) inhibition by demethylating agents. (**A**) Inhibition of DNMTs by 5-Aza-CdR (or 5-Aza-CR). After incorporation of the analog into genomic DNA, the aza-substituted DNA forms an irreversible covalent complex with the DNMT, thereby depleting the enzyme and resulting in passive demethylation over successive rounds of DNA replication. This mechanism is representative of other nucleoside analog inhibitors of DNA methylation. (**B**) Inhibition of DNMTs by antisense inhibitors. The addition of a short antisense oligomer leads to the formation of a region of double stranded RNA (dsRNA) that is degraded and subsequently prevents translation of the DNMT protein, leading to DNA demethylation. This mechanism of inhibition is similar to that of small-interfering RNA.

(Bouchard and Momparler, 1983; Santi, Garrett and Barr, 1983). The trapping and subsequent depletion of the DNMT enzyme results in a passive demethylation process in which newly synthesized DNA strands remain hypomethylated and the loss of methylation patterns is propagated during cellular replication (Bender et al., 1999; Jones and Taylor, 1981; Jones, Taylor and Wilson, 1983; Velicescu et al., 2002; Wilson, Jones and Momparler, 1983).

Demethylation by these agents has shown to result in the reexpression of numerous methylation-silenced genes, such as *p15*, *p16*, *BRCA1*, *RB*, *hMLH1*, and *Myf3* (Bender, Zingg and Jones, 1998b; Jones, 1985; Strathdee et al., 1999). Furthermore, 5-Aza-CdR has been shown to induce cellular differentiation (Constantinides, Jones and Gevers, 1977; Jones and Taylor, 1980) and has been used in clinical trials for the treatment of several hematopoietic disorders (Pinto and Zagonel, 1993; Issa et al., 2004). At equimolar concentrations, 5-Aza-CdR is at least twice as potent as 5-Aza-CR in inhibiting DNA methylation (Pinto and Zagonel, 1993), since 5-Aza-CR is predominantly incorporated into newly synthesized RNA in addition to DNA, whereas 5-Aza-CdR is incorporated exclusively into DNA. The incorporation of 5-Aza-CR into RNA produces disassembly of polyribosomes, defective methylation and acceptor function of transfer RNA, and marked inhibition of protein production (Glazer and Hartman, 1980; Lu et al., 1976). Both drugs are inactivated through deamination by cytidine deaminase. The clinical utility of these agents is limited by their cytotoxicities at high concentration and their instability in aqueous solutions, making frequent changes of newly prepared solutions necessary (Issa et al., 2003; Jones and Baylin, 2002; Santini, Kantarjian and Issa, 2001).

In addition to 5-Aza-CR and 5-Aza-CdR, various cytosine analogs modified in position 5 of the pyrimidine ring, such as pseudoisocytidine and 5-fluoro-2-deoxy-cytidine, have also been developed and shown to be potent inhibitors of DNA methylation (Figure 12.2) (Jones and Taylor, 1980). Other molecular variations of 5-Aza-CR were developed, including dihydro-5-azacytidine (DHAC) (a hydrolyti-cally stable congener of 5-Aza-CR) and arabinofuranosyl-5-azacytosine (fazara-bine), both of which were shown to be effective inhibitors of DNA methylation and introduced into clinical trials (Figure 12.2) (Curt et al., 1985; Glazer and Knode, 1984; Goldberg et al., 1997; Powell and Avramis, 1988; Samuels et al., 1998; Yogelzang et al., 1997). This distinctive feature of the modification at position C-5 is seemingly responsible for inhibiting DNMT. Analogs that do not possess this change in the pyrimidine ring, such as cytosine arabinoside (Ara-CR) (Figure 12.2), 6-azacytidine (6-Aza-CR), and gemcitabine, do not inhibit DNA methylation (Jones and Taylor, 1980; Santini et al., 2001). However, nucleoside analog inhibitors of DNA methyltransferase, such as aza-nucleoside analogs, are limited somewhat by its myelosuppression in patients and potential mutagenic effects following the incor-poration of these agents into the genomic DNA (Hsiao, Gattoni-Celli and Weinstein, 1985; Jackson-Grusby et al., 1997). Unfortunately, despite promising results with these agents in various clinical trials, the mutagenic, cytotoxic, and chemically unstable properties of nucleoside analogs have limited their clinical effectiveness.

Aza-nucleoside analogs are cytotoxic and myelosuppresssive, especially when used at the high doses commonly used in clinical trials. On the other hand, these agents can reactivate silenced genes and cause cellular differentiation at low doses. The efficiency of gene reactivation by 5-aza analogs is not necessarily improved by increasing the dosage beyond their optimally effective concentration (Jones and Taylor, 1980). As a matter of fact, a recent study has shown that low doses may be more efficacious in myeloid malignancies, with greater therapeutic effects and better tolerance and side effect profiles, than high doses (Issa et al., 2003). This idea of finding the optimal dose for responses rather than a maximally tolerated dose will definitely change the way our future clinical trials are conducted.

Interestingly, a nucleoside analog without the apparent modification at position 5 of pyrimidine ring, zebularine (1-(beta-D-ribofuranosyl)-1,2-dihydropyrimidin-2-one), was recently characterized as a novel inhibitor of DNA methylation (Figure 12.2) (Cheng et al., 2003). Zebularine, a cytidine analog containing a 2-(1H)-pyrimidinone ring, was originally developed as a cytidine deaminase inhibitor because it lacks an amino group on position 4 of the ring (Kim et al., 1986; Laliberte, Marquez and Momparler, 1992) and is known to be stable in acidic and neutral aqueous solutions and minimally toxic *in vitro* and *in vivo* (Cheng et al., 2003). The mechanism of action is presumably similar to 5-Aza-CR, which requires the incorporation into DNA and subsequent formation of a covalent bond with the DNMT (Hurd et al., 1999; Zhou et al., 2002). Due to its chemical stability, this agent was the first in its class shown to be orally active in the reactivation of genes silenced by aberrant methylation in mouse models (Cheng et al., 2003), however, this drug has yet to be tested in clinical trials. Nevertheless, with its stability and minimal toxicity, zebularine holds promise as an effective anticancer agent.

Nonnucleoside Analogs

Other than nucleoside analogs, nonnucleoside analogs have also been described to function as inhibitors of DNA methylation. These inhibitors are of interest because they circumvent a number of the potential drawbacks associated with nucleoside analog treatments, including the requirement for DNA incorporation and the formation of toxic protein–DNA adducts. The first of these inhibitors described were antisense oligonucleotides directed against the DNMT1 mRNA (Fournel et al., 1999; MacLeod and Szyf, 1995; Ramchandani et al., 1997). Antisense oligonucleotide inhibitors are relatively short synthetic nucleic acids designed to hybridize to a specific mRNA sequence, and the hybridization can block mRNA translation and cause mRNA degradation (Figure 12.3B) (Crooke, 2000). Antisense oligonucleotides directed against DNMT1 mRNA have been shown to inhibit DNMT1 mRNA expression, DNMT1 activity, and tumorigenesis *ex vivo* and *in vivo* (Fournel et al., 1999; MacLeod and Szyf, 1995; Ramchandani et al., 1997). Since many of these agents are associated with nonspecific toxicities (Akhtar and Agrawal, 1997), a second generation DNMT1 antisense bearing an O-methyl modification of the ribose in the four 5' and 3' nucleotides was generated (Fournel et al., 1999). This modified antisense effectively inhibited DNMT1 and exhibited very strong antitumor effects *in vitro* and *in vivo*. DNMT1 antisense (MG98) oligonucleotides are currently in Phase 1 clinical trials. Along the same line, small-interfering RNAs (siRNAs) of DNMT1 and DNMT3b were recently developed and demonstrated to result in selective depletion of these enzymes with corresponding lower methyltransferase activities and global and gene-specific demethylation and reactivation of tumor suppressor genes in human cancer cells (Leu et al., 2003; Matsukura, Jones and Takai, 2003; Robert et al., 2003).

A second novel method for inhibiting DNMT1 consists of hairpin-structured oligonucleotide substrate mimics. These inhibitors act competitively and inhibit purified DNMT1 mRNA at nanomolar concentrations *in vitro* (Bigey et al., 1999). However, for reasons that are unclear, the hairpin inhibitors of DNMT1 are unable to induce DNA methylation changes or activate methylation-silenced genes in treated cells (Bigey et al., 1999). Nonetheless, antisense oligonucleotide inhibitors remain promising candidates for anticancer therapy.

Other nonnucleoside analogs that inhibit methylation include procainamide, which is used for the treatment of cardiac arrhythmias, and hydralazine, which is used for the treatment of hypertension. Procainamide is a noncompetitive inhibitor of the DNA methyltransferases and can reactivate the transcription of the *p16* gene from a methylation-silenced promoter in prostate cancer cells grown in nude mice (Lin et al., 2001). Both agents have been shown to induce deme-thylation and reexpression of estrogen receptor (*ER*), retinoic acid receptor β (*RAR-β*), and *p16* genes in cultured cells, and the *ER* gene in mice. Oral treatment with hydralazine was also able to demethylate and reexpress the *RAR-β* and *p16* genes in cancer patients (Segura-Pacheco et al., 2003). In addition, an anesthetic drug related to procainamide, procaine, was also found to effectively inhibit DNA methylation in breast cancer cells (Villar-Garea et al., 2003). The mechanism by which these three agents (procainamide, hydralazine, and procaine) inhibit DNA methylation is still unknown, however, it has been postulated that these drugs bind to CG-rich DNA sequences (Thomas and Messner, 1986; Villar-Garea et al., 2003), thereby interfering with the translocation of DNA methyltransferase along the DNA strands. Although these agents may have various non-specific medical side effects unrelated to their demethylating effects (Knowles, Shapiro and Shear, 2003), their availabilities for oral use, low costs, and negligible toxicities make them suitable for clinical trials and offer a simple alternative to 5-Aza-CdR and antisense oligonuleotides.

S-ADENOSYLMETHIONINE (ADOMET)-RELATED INHIBITORS OF DNA METHYLATION

DNA methylation can also be inhibited by targeting the pathways of S-adenosyl-L-methionine (SAM) metabolism. Many of these agents that inhibit DNA methyl-ation are competitive inhibitors of SAM, inhibitors of SAM synthesis, or inhibitors of S-adenosyl-L-homocysteine (SAH) metabolism to target SAM-dependent meth-yltransferases (Bender et al., 1998b). Numerous studies have demonstrated the inhibition of methyl-transfer and decarboxylation reactions using analogs and metabolites of SAM (Borchardt, 1980; Pegg, Secrist and Madhubala, 1988). There are many ways that SAM metabolism can be altered, however, the interference with SAM metabolism may lead to nonspecific effects on other cellular methyl-transfer reactions that require SAM, including spermidine and spermine synthesis (Chiang et al., 1996). This, therefore, limits the clinical utility of these compounds that interfere with all SAM-dependent methyl-transfer reactions. Furthermore, many of these inhibitors of (cytosine-5)-DNA methyltransferases analogous to SAM and SAH have been shown to be mutagenic (Zingg et al., 1996).

Given the therapeutic promise of reversing methylation changes in cancer, there are certain to be more DNA methylation inhibitors for use in clinical studies. The ultimate goal is to identify or develop an inhibitor that exhibits greatest demethy-lating ability, yet with minimal side effects and drawbacks.

DRAWBACKS OF DEMETHYLATING AGENTS AND CLINICAL CONCERNS

INAPPROPRIATE ACTIVATION OF GENES

A major concern regarding the clinical use of demethylating agents is the potential inappropriate activation of genes in normal cells. This may not, however, be that serious, since transcription from promoter-containing CpG islands does not seem to be commonly controlled by methylation in normal cells. There is little evidence indicating that demethylation causes substantial changes in gene expression in normal cells (Liang et al., 2002), where the major targets would be imprinted genes and X-chromosome inactivated genes. Demethylation of imprinted genes may not be a critical issue in most adult cells. In addition, the process of X-chromosome inactivation is also known to be most important during embryogenesis and modulated by different layers of epigenetic control, such as RNA expression, histone modification, and chromatin remodeling (Bird and Wolffe, 1999). In fact, studies have shown that the X-chromosome cannot be reactivated in normal human fibroblasts after treatment with 5-Aza-CdR (Wolf and Migeon, 1982), whereas it can be readily reactivated in rodent–human somatic cells (Wolf and Migeon, 1982), indicating that some higher level of control is likely to be absent in these hybrids. Apparently, both imprinted genes and X-chromosome inactivated genes cannot be routinely and easily activated by inhibiting DNA methylation in normal adult cells. Besides these genes, another important concern involves the potential activation of transposable elements, which are normally methylated, and the consequential deleterious effects in normal tissues. The activation of these transposable elements by demethylating agents is still unclear and thus requires further studies to better understand this phenomenon. However, there are encouraging results indicating that patients treated with 5-Aza-nucleosides for various malignant and nonmalignant diseases did not generally show massive toxicity due to inappropriate gene activation (Lübbert, 2000).

REMETHYLATION AND RESILENCING

Another impediment to the use of demethylation therapy is the phenomenon of gene resilencing. Following the removal of the demethylating agent, there appears to be a tendency of methylation to spread back into the demethylated CpG island in these cells (Jones and Baylin, 2002). This process has been shown to lead to *de novo* methylation and to the resilencing of the target gene (Bender et al., 1999). In fact, transfected retroviral long terminal repeat (LTR) promoters have been shown to be actively methylated in specific cancer cell lines (Lengauer, Kinzler and Vogelstein, 1997). This short therapeutic window due to rapid gene silencing definitely poses critical challenges for pharmacological strategies that seek to reactivate tumor suppressor gene expression. Despite this challenge, this narrow window of demethylation can be used for appropriate scheduling of secondary agents whose antitumor effects are being nullified due to abnormal gene hypermethylation. In fact, promising results from our laboratory have shown that zebularine can hinder or prevent remethylation or resilencing from occurring when the drug is administered in a continuous fashion to cancer cells (Cheng et al., 2004). The potential clinical application of this drug remains to be elucidated.

NONSPECIFICITY

The issue of nonspecificity of demethylating agents is also a potential drawback. Demethylating agents have also been shown to activate genes that either do not have CpG islands in their promoters (Liang et al., 2002) or are unmethylated to begin with (Soengas et al., 2001; Suzuki et al., 2002), and there is insufficient information with regard to the spectrum of genes that are activated in normal cells after exposure to such agents. The nonspecific effects of demethylating agents on both normal and cancer cells will limit their clinical utility. The pharmacological or biological approaches in the inhibition of cytosine methylation, however, have nevertheless been promising in preventing or reversing the onset neoplasia as shown in various animal models (Belinsky et al., 2003; Laird et al., 1995; Lantry et al., 1999). Furthermore, various neoplastic cell lines were shown to be growth suppressed after the treatment with 5-Aza-CdR, whereas normal fibroblasts cell lines were unaffected (Bender, Pao and Jones, 1998a). These studies along with others have renewed interest in the clinical application of demethylating agents, though there is still a great deal of the promise that remains to be fulfilled.

REACTIVATION OF GENE EXPRESSION THROUGH DRUG COMBINATIONS

The reactivation of silenced genes by DNA methylation inhibitors can result in the direct suppression of tumor growth or sensitization of cancer cells to other anticancer therapies. An exciting approach in the application of demethylating agents includes the combination treatment with a demethylating agent initially, followed by a secondary agent that targets the reactivated genes or pathway. The combination of a demethylating agent with a secondary agent allows a more distinct and direct therapeutic approach in targeting specific genes or reactivated pathways in the treatment of human cancer.

DEMETHYLATING AGENT AND CHEMOTHERAPY

There are several examples illustrating the usage of demethylating agent in reactivating critical cellular processes to sensitize cancer cells resistant to chemotherapy treatment (Table 12.1). Many chemotherapeutic agents are known to kill susceptible cells through the induction of the physiological cell death program (apoptosis). Apparently, deregulation of any gene involved in the activation or execution of apoptosis may be an important mechanism of chemoresistance (Schmitt and Lowe, 1999). A prominent example of this includes the association between loss of apoptosis-related protein *caspase-8* and resistance to cytotoxic drugs like doxorubicin and cisplatin. Treatment of cells harboring *caspase-8* promoter hypermethylation with 5-Aza-CdR led to reexpression of *caspase-8* and restored sensitivity for chemotherapy,

TABLE 12.1
Drug Combination Treatments with Demethylating Agents

Secondary Agents	Reactivated Genes Pathway	Results or Possible Goals	References
Established Drug Combinations			
IFN-α	*STAT1, 2, 3* and interferon-related genes/IFN signal transduction pathway	Sensitize IFN-resistant tumor	Karpf et al., 1999
Cisplatin Carboplatin Temozolomide Epirubicin	*MLH1*/Mismatch repair pathway	Sensitize xenografts resistant to chemotherapy due to the loss of mismatch repair	Plumb et al., 2000
Adriamycin	*Apaf-1*/Apoptotic pathway	Enhance sensitivity to adriamycin and a rescue of the apoptotic defects associated with *Apaf-1* silencing	Soengas et al., 2001
HDAC inhibitors (Trichostatin A, Phenylbutyrate)	Various genes/Pathways	Reactivate gene expression synergistically in cancer cells	Cameron et al., 1999 Belinsky et al., 2003
Doxorubicin Cisplatin	*Caspase-8*/Apoptotic pathway	Restore sensitivity for chemotherapy or death-receptor-induced apoptosis in various cancers	Zheng et al., 2000 Fulda et al., 2001
Potential Drug Combinations			
Retinoic acid	*RAR-β*/Retinoic acid signaling pathway	Reconstitution of retinoic acid signaling and the responsiveness of cancer cells to retinoids	Cote et al., 1998 Yang et al., 2002
Immunotherapy	*MAGE1*/Cancer-related antigens	Upregulation of cancer-related antigens for targeting for immunotherapy	Weber et al., 1994
Antiestrogen/endocrine therapy	*ER*/ER pathway	Sensitize ER-negative breast cancer cells to endocrine therapy	Lapidus et al., 1998

or death-receptor-induced apoptosis in various tumors *in vivo* (Teitz et al., 2000). The resistance to chemotherapeutic agents appears to be mediated by methylation of genes in the apoptotic pathway.

In another instance, the restoration of *Apaf-1* expression in highly chemoresistant melanoma cell lines after treatment with 5-Aza-CdR led to a marked enhancement in their sensitivity to adriamycin and a rescue of the apoptotic defects associated with *Apaf-1* silencing (Soengas et al., 2001). 5-Aza-CdR treatment was also shown to be effective in sensitizing previously resistant ovarian cancer cells or tumor xenografts to cisplatin treatment (Plumb et al., 2000; Strathdee et al., 1999). The mechanistic basis for the sensitization to cisplatin is presumably due to the reactivation of the methylation-silenced *MLH1* mismatch repair gene. Altogether, these studies support the idea of targeting the silencing mechanism of genes involved in critical cellular processes, such as apoptosis or DNA repair, with demethylating agents to increase the efficacy of various forms of chemotherapy.

DEMETHYLATING AGENT AND HISTONE DEACETYLASE INHIBITORS

An interesting combination treatment which has received interest recently is the combination usage of DNA demethylating agents with histone deacetylase (HDAC) inhibitors (Table 12.1). Histone modification and chromatin remodeling are also known as major epigenetic players in the transcriptional regulation of gene expression (Grewal and Moazed, 2003; Turner, 2002). HDAC activity has been shown to be important in the transcriptional repression of methylated sequences (Brown and Strathdee, 2002). Previously, it was also shown that combination of 5-Aza-CdR and trichostatin A, an inhibitor of HDACs, causes synergistic reactivation of gene expression in cancer cell lines (Cameron et al., 1999). Another recent study also showed that combination of 5-Aza-CdR and the HDAC inhibitor sodium phenylbutyrate can synergistically reactivate gene expression in human cancer cells and effectively reduce formation of lung cancers in mouse models (Belinsky et al., 2003). Given the cross-talk between these epigenetic pathways, combinations of these drugs could be more effective than individual drugs alone. Their synergistic effect on gene expression can possibly prolong the timing of the therapeutic window, making this a powerful combination regimen for reactivating silenced genes.

DEMETHYLATING AGENT AND INTERFERON

The strategy of using a demethylating agent together with a secondary agent has also been demonstrated with the interferon pathway (Table 12.1). Treatment of various cancer cell lines with 5-Aza-CdR leads to an increased expression of genes involved in the interferon signal transduction pathway (Karpf and Jones, 2002; Karpf et al., 1999; Liang et al., 2002). The induction of interferon pathway target genes following the 5-Aza-CdR treatment was most likely an indirect effect of the drug, since most of the interferon-related genes do not contain promoter CpG islands. Nevertheless, these findings have strong clinical implications, since it has been previously reported that interferon responsive genes become down-regulated in various tumors as compared to their normal counterparts (Karpf et al., 1999), and that the interferon pathway may be disrupted during carcinogenesis (Abril et al., 1998; Sun et al., 1998; Wong et al., 1997). In fact, treatment with 5-Aza-CdR was shown to sensitize cancer cells to subsequent treatment with interferon-α (Karpf et al., 1999). It will thus be of great interest to assess the ability of demethylating agents to reconstitute interferon sensitivity in cancer patients who are nonresponsive to interferon-α therapy.

POTENTIAL COMBINATION THERAPIES

Other potential combination strategies have been proposed besides these combination treatments (Table 12.1). Potential secondary agents that have been suggested include (1) estrogen therapy, (2) retinoids, and (3) immunotherapy. First, methylation of the ER promoter CpG island appears to play a role in its silencing in ER-negative tumor cell lines (Lapidus et al., 1998). Given the fact that 17β-estradiol stimulates the growth of ER-positive breast tumors via functional ER, endocrine therapy such as antiestrogen or ovarian ablation has been established as an important part of breast cancer management (Davidson, 2001). Thus, it may be feasible to reactivate the ER protein and sensitize patients with ER-negative tumors to become responsive to endocrine therapy. Second, the *RAR-β* gene was shown to be silenced by DNA methylation in numerous cancers (Cote, Sinnett and Momparler, 1998; Yang et al., 2002). It will thus be interesting to evaluate whether the reactivation of the *RAR-β* gene by demethylation can reconstitute the responsiveness of various cancer cells to retinoids (Karpf and Jones, 2002). Last, studies have shown that 5-Aza-CdR treatment can upregulate the expression of specific MAGE antigens, a family of cancer/testis-specific antigens (Weber et al., 1994). Apparently, this MAGE gene induction is specific to tumor and not normal epithelial cells (Karpf and Jones, 2002). These data suggest a potential combination treatment that involves the use of demethylating agents to augment the presentation of cancer-specific antigens, which can thereby be targeted by immunotherapy.

CONCLUSION

The prevalence of aberrant epigenetic inactivation of critical genes in tumors makes an attractive target for inhibitors of DNA methylation as novel anticancer therapies. The search and identification of newer demethylating agents that are both effective biologically effective and clinically beneficial without the various drawbacks will inevitably continue. The search for an ideal demethylating agent that exhibits potent demethylating abilities without drawbacks is highly unlikely; however, the agents with greatest effects or benefits and fewest side effects would be the most practical and relevant in the clinical setting. Furthermore, the diverse possibilities in the combined application of demethylating agents with secondary agents offer various new exciting clinical strategies that can specifically target reactivated genes and pathways in the treatment of cancer.

REFERENCES

Abril, E., Real, L. M., Serrano, A., Jimenez, P., Garcia, A., Canton, J., Trigo, I., Garrido, F. and Ruiz-Cabello, F., Unresponsiveness to interferon associated with STAT1 protein deficiency in a gastric adenocarcinoma cell line, *Cancer Immunol. Immunother.*, 47, 113–120, 1998.

Akhtar, S. and Agrawal, S., In vivo studies with antisense oligonucleotides. *Trends Pharmacol. Sci.*, 18, 12–18, 1997.

Belinsky, S. A., Klinge, D. M., Stidley, C. A., Issa, J. P., Herman, J. G., March, T. H. and Baylin, S. B., Inhibition of DNA methylation and histone deacetylation prevents murine lung cancer, *Cancer Res.*, 63, 7089–7093, 2003.

Bender, C. M., Gonzalgo, M. L., Gonzales, F. A., Nguyen, C. T., Robertson, K. D. and Jones, P. A., Roles of cell division and gene transcription in the methylation of CpG islands, *Mol. Cell. Biol.*, 19, 6690–6698, 1999.

Bender, C. M., Pao, M. M. and Jones, P. A., Inhibition of DNA methylation by 5-aza-2-deoxycytidine suppresses the growth of human tumor cell lines, *Cancer Res.*, 58, 95–101, 1998a.

Bender, C. M., Zingg, J. M. and Jones, P. A., DNA methylation as a target for drug design, *Pharm Res.*, 15, 175–187, 1998b.

Bigey, P., Knox, J. D., Croteau, S., Bhattacharya, S. K., Theberge, J. and Szyf, M., Modified oligonucleotides as bona fide antagonists of proteins interacting with DNA. Hairpin antagonists of the human DNA methyltransferase, *J. Biol. Chem.*, 274, 4594–4606, 1999.

Bird, A. P. and Wolffe, A. P., Methylation-induced repression — belts, braces, and chromatin, *Cell*, 99, 451–454, 1999.

Borchardt, R. T., S-Adenosyl-L-methionine-dependent macromolecule methyltransferases: potential targets for the design of chemotherapeutic agents, *J. Med. Chem.*, 23, 347–357, 1980.

Bouchard, J. and Momparler, R. L., Incorporation of 5-aza-2′-deoxycytidine-5′-triphosphate into DNA. Interactions with mammalian DNA polymerase alpha and DNA methylase, *Mol. Pharmacol.*, 24, 109–114, 1983.

Brown, R. and Strathdee, G., Epigenomics and epigenetic therapy of cancer, *Trends Mol. Med.*, 8, S43–S48, 2002.

Cameron, E. E., Bachman, K. E., Myohanen, S., Herman, J. G. and Baylin, S. B., Synergy of demethylation and histone deacetylase inhibition in the re-expression of genes silenced in cancer, *Nat. Genet.*, 21, 103–107, 1999.

Cheng, J. C., Weisenberger, D. J., Gonzales, F. A., Liang, G., Xu, G. L., Hu, Y. G., Marquez, V. E. and Jones, P. A., Continuous zebularine treatment effectively sustains demethylation in human bladder cancer cells, *Mol. Cell. Biol.*, 24, 1270–1278, 2004.

Cheng, J. C., Matsen, C. B., Gonzales, F. A., Ye, W., Greer, S., Marquez, V. E., Jones, P. A. and Selker, E. U., Inhibition of DNA methylation and reactivation of silenced genes by zebularine, *J. Natl. Cancer Inst.*, 95, 399–409, 2003.

Chiang, P. K., Gordon, R. K., Tal, J., Zeng, G. C., Doctor, B. P., Pardhasaradhi, K. and McCann, P. P., S-Adenosylmethionine and methylation, *Faseb. J.*, 10, 471–480, 1996.

Constantinides, P. G., Jones, P. A. and Gevers, W., Functional striated muscle cells from non-myoblast precursors following 5-azacytidine treatment, *Nature*, 267, 364–366, 1997.

Cote, S., Sinnett, D. and Momparler, R. L., Demethylation by 5-aza-2'-deoxycytidine of specific 5-methylcytosine sites in the promoter region of the retinoic acid receptor beta gene in human colon carcinoma cells, *Anticancer Drug*, 9, 743–750, 1998.

Crooke, S. T., Potential roles of antisense technology in cancer chemotherapy. *Oncogene*, 19, 6651–6659, 2000.

Curt, G. A., Kelley, J. A., Fine, R. L., Huguenin, P. N., Roth, J. S., Batist, G., Jenkins, J. and Collins, J. M., A phase I and pharmacokinetic study of dihydro-5-azacytidine (NSC 264880), *Cancer Res.*, 45, 3359–3363, 1985.

Davidson, N. E., Ovarian ablation as adjuvant therapy for breast cancer, *J. Natl. Cancer Inst. Monogr.*, 67-71, 2001.

Fournel, M., Sapieha, P., Beaulieu, N., Besterman, J. M. and MacLeod, A. R., Down-regulation of human DNA-(cytosine-5) methyltransferase induces cell cycle regulators p16(ink4A) and p21(WAF/Cip1) by distinct mechanisms, *J. Biol. Chem.*, 274, 24250–24256, 1999.

Fulda, S., Kufer, M. U., Meyer, E., van Valen, F., Dockhorn-Dworniczak, B. and Debatin, K. M., Sensitization for death-receptor – or drug-induced apoptosis by re-expression of caspase-8 through demethylation or gene transfer, *Oncogene*, 20, 5865–5877, 2001.

Glazer, R. I. and Hartman, K. D., The comparative effects of 5-azacytidine and dihydro-5-azacytidine on 4 S and 5 S nuclear RNA, *Mol. Pharmacol.*, 17, 250–255, 1980.

Glazer, R. I. and Knode, M. C., 1-Beta-D-arabinosyl-5-azacytosine. Cytocidal activity and effects on the synthesis and methylation of DNA in human colon carcinoma cells, *Mol. Pharmacol.*, 26, 381–387, 1984.

Goldberg, R. M., Reid, J. M., Ames, M. M., Sloan, J. A., Rubin, J., Erlichman, C., Kuffel, M. J. and Fitch, T. R., Phase I and pharmacological trial of fazarabine (Ara-AC) with granulocyte colony-stimulating factor, *Clin. Cancer Res.*, 3, 2363–2370, 1997.

Grewal, S. I. and Moazed, D., Heterochromatin and epigenetic control of gene expression, *Science*, 301, 798–802, 2003.

Hsiao, W. L., Gattoni-Celli, S. and Weinstein, I. B., Effects of 5-azacytidine on the progressive nature of cell transformation, *Mol. Cell. Biol.*, 5, 1800–1803, 1985.

Hurd, P. J., Whitmarsh, A. J., Baldwin, G. S., Kelly, S. M., Waltho, J. P., Price, N. C., Connolly, B. A. and Hornby, D. P., Mechanism-based inhibition of C5-cytosine DNA methyltransferases by 2-H pyrimidinone, *J. Mol. Biol.*, 286, 389–401, 1999.

Issa, J. P., Garcia-Manero, G., Giles, F. J., Mannari, R., Thomas, D., Faderl, S., Bayar, E., Lyons, J., Rosenfeld, C. S., Cortes, J. and Kantarjian, H. M., Phase I study of low-dose prolonged exposure schedules of the hypomethylating agent 5-aza-2'-deoxycytidine (Decitabine) in hematopoietic malignancies, *Blood*, 103, 1635–1640, 2004.

Jackson-Grusby, L., Laird, P. W., Magge, S. N., Moeller, B. J. and Jaenisch, R., Mutagenicity of 5-aza-2'-deoxycytidine is mediated by the mammalian DNA methyltransferase, *Proc. Natl. Acad. Sci. USA*, 94, 4681–4685, 1997.

Jones, P. A., Effects of 5-azacytidine and its 2'-deoxyderivative on cell differentiation and DNA methylation, *Pharmacol. Ther.*, 28, 17–27, 1985.

Jones, P. A. and Baylin, S. B., The fundamental role of epigenetic events in cancer, *Nat. Rev. Genet.*, 3, 415–428, 2002.

Jones, P. A. and Taylor, S. M., Cellular differentiation, cytidine analogs and DNA methylation, *Cell*, 20, 85–93, 1980.

Jones, P. A. and Taylor, S. M., Hemimethylated duplex DNAs prepared from 5-azacytidine-treated cells, *Nucleic Acids Res.*, 9, 2933–2947, 1981.

Jones, P. A., Taylor, S. M. and Wilson, V. L., Inhibition of DNA methylation by 5-azacytidine, *Recent Results Cancer Res.*, 84, 202–211, 1983.

Karpf, A. R. and Jones, D. A., Reactivating the expression of methylation silenced genes in human cancer, *Oncogene,* 21, 5496–5503, 2002.

Karpf, A. R., Peterson, P. W., Rawlins, J. T., Dalley, B. K., Yang, Q., Albertsen, H. and Jones, D. A., Inhibition of DNA methyltransferase stimulates the expression of signal transducer and activator of transcription 1, 2, and 3 genes in colon tumor cells, *Proc. Natl. Acad. Sci. USA,* 96, 14007–14012, 1999.

Kim, C. H., Marquez, V. E., Mao, D. T., Haines, D. R. and McCormack, J. J., Synthesis of pyrimidin-2-one nucleosides as acid-stable inhibitors of cytidine deaminase, *J. Med. Chem.,* 29, 1374–1380, 1986.

Knowles, S. R., Shapiro, L. E. and Shear, N. H., Reactive metabolites and adverse drug reactions: clinical considerations, *Clin. Rev. Allergy Immunol.,* 24, 229–238, 2003.

Laird, P. W., Jackson-Grusby, L., Fazeli, A., Dickinson, S. L., Jung, W. E., Li, E., Weinberg, R. A. and Jaenisch, R., Suppression of intestinal neoplasia by DNA hypomethylation, *Cell,* 81, 197–205, 1995.

Laliberte, J., Marquez, V. E. and Momparler, R. L., Potent inhibitors for the deamination of cytosine arabinoside and 5-aza-2′-deoxycytidine by human cytidine deaminase, *Cancer Chemother. Pharmacol.,* 30, 7–11, 1992.

Lantry, L. E., Zhang, Z., Crist, K. A., Wang, Y., Kelloff, G. J., Lubet, R. A. and You, M., 5-Aza-2′-deoxycytidine is chemopreventive in a 4-(methyl-nitrosamino)-1-(3-pyridyl)-1-butanone-induced primary mouse lung tumor model, *Carcinogenesis,* 20, 343–346, 1999.

Lapidus, R. G., Nass, S. J., Butash, K. A., Parl, F. F., Weitzman, S. A., Graff, J. G., Herman, J. G. and Davidson, N. E., Mapping of ER gene CpG island methylation-specific polymerase chain reaction, *Cancer Res.,* 58, 2515–2519, 1998.

Lengauer, C., Kinzler, K. W. and Vogelstein, B., DNA methylation and genetic instability in colorectal cancer cells,. *Proc. Natl. Acad. Sci. USA,* 94, 2545–2550, 1997.

Leu, Y. W., Rahmatpanah, F., Shi, H., Wei, S. H., Liu, J. C., Yan, P. S. and Huang, T. H., Double RNA interference of DNMT3b and DNMT1 enhances DNA demethylation and gene reactivation, *Cancer Res.,* 63, 6110–6115, 2003.

Liang, G., Gonzales, F. A., Jones, P. A., Orntoft, T. F. and Thykjaer, T., Analysis of gene induction in human fibroblasts and bladder cancer cells exposed to the methylation inhibitor 5-aza-2′-deoxycytidine, *Cancer Res.,* 62, 961–966, 2002.

Lin, X., Asgari, K., Putzi, M. J., Gage, W. R., Yu, X., Cornblatt, B. S., Kumar, A., Piantadosi, S., DeWeese, T. L., De Marzo, A. M. and Nelson, W. G., Reversal of GSTP1 CpG island hypermethylation and reactivation of pi-class glutathione S-transferase (GSTP1) expression in human prostate cancer cells by treatment with procainamide, *Cancer Res.,* 61, 8611–8616, 2002.

Lu, L. W., Chiang, G. H., Medina, D. and Randerath, K., Drug effects on nucleic acid modification. I. A specific effect of 5-azacytidine on mammalian transfer RNA methylation in vivo, *Biochem. Biophys. Res. Commun.,* 68, 1094–1001, 1976.

Lübbert, M., DNA methylation inhibitors in the treatment of leukemias, myelodysplastic syndromes and hemoglobinopathies: clinical results and possible mechanisms of action, *Curr. Top. Microbiol. Immunol.,* 249, 135–164, 2000.

MacLeod, A. R. and Szyf, M., Expression of antisense to DNA methyltransferase mRNA induces DNA demethylation and inhibits tumorigenesis, *J. Biol. Chem.,* 270, 8037–8043, 1995.

Matsukura, S., Jones, P. A. and Takai, D., Establishment of conditional vectors for hairpin siRNA knockdowns, *Nucleic Acids Res.,* 31, e77, 2003.

Pegg, A. E., Secrist III, J. A. and Madhubala, R., Properties of L1210 cells resistant to alpha-difluoromethylornithine, *Cancer Res.,* 48, 2678–2682, 1988.

Pinto, A. and Zagonel, V., 5-Aza-2'-deoxycytidine (Decitabine) and 5-azacytidine in the treatment of acute myeloid leukemias and myelodysplastic syndromes: past, present and future trends, *Leukemia*, 7, Suppl 1, 51–60, 1993.

Plumb, J. A., Strathdee, G., Sludden, J., Kaye, S. B. and Brown, R., Reversal of drug resistance in human tumor xenografts by 2'-deoxy-5-azacytidine-induced demethylation of the hMLH1 gene promoter, *Cancer Res.*, 60, 6039–6044, 2000.

Powell, W. C. and Avramis, V. I., Biochemical pharmacology of 5,6-dihydro-5-azacytidine (DHAC) and DNA hypomethylation in tumor (L1210)-bearing mice, *Cancer Chemother. Pharmacol.*, 21, 117–121, 1988.

Ramchandani, S., MacLeod, A. R., Pinard, M., von Hofe, E. and Szyf, M., Inhibition of tumorigenesis by a cytosine-DNA, methyltransferase, antisense oligodeoxynucleotide, *Proc. Natl. Acad. Sci. USA*, 94, 684–689, 1997.

Robert, M. F., Morin, S., Beaulieu, N., Gauthier, F., Chute, I. C., Barsalou, A. and MacLeod, A. R., DNMT1 is required to maintain CpG methylation and aberrant gene silencing in human cancer cells, *Nat. Genet.*, 33, 61–65, 2003.

Samuels, B. L., Herndon, J. E., 2nd, Harmon, D. C., Carey, R., Aisner, J., Corson, J. M., Suzuki, Y., Green, M. R. and Vogelzang, N. J., Dihydro-5-azacytidine and cisplatin in the treatment of malignant mesothelioma: a phase II study by the Cancer and Leukemia Group B, *Cancer*, 82, 1578–1584, 1988.

Santi, D. V., Garrett, C. E. and Barr, P. J., On the mechanism of inhibition of DNA-cytosine methyltransferases by cytosine analogs, *Cell*, 33, 9–10, 1983.

Santini, V., Kantarjian, H. M. and Issa, J. P., Changes in DNA methylation in neoplasia: pathophysiology and therapeutic implications, *Ann. Intern. Med.*, 134, 573–586, 2001.

Schmitt, C. A. and Lowe, S. W., Apoptosis and therapy. *J. Pathol.*, 187, 127–137, 1999.

Segura-Pacheco, B., Trejo-Becerril, C., Perez-Cardenas, E., Taja-Chayeb, L., Mariscal, I., Chavez, A., Acuna, C., Salazar, A. M., Lizano, M. and Duenas-Gonzalez, A., Reactivation of tumor suppressor genes by the cardiovascular drugs hydralazine and procainamide and their potential use in cancer therapy, *Clin. Cancer Res.*, 9, 1596–1603, 2003.

Soengas, M. S., Capodieci, P., Polsky, D., Mora, J., Esteller, M., Opitz-Araya, X., McCombie, R., Herman, J. G., Gerald, W. L., Lazebnik, Y. A., Cordon-Cardo, C. and Lowe, S. W., Inactivation of the apoptosis effector Apaf-1 in malignant melanoma, *Nature*, 409, 207–211, 2001.

Sorm, F., Piskala, A., Cihak, A. and Vesely, J., 5-Azacytidine, a new, highly effective cancerostatic, *Experientia*, 20, 202–203, 1964

Strathdee, G., MacKean, M. J., Illand, M. and Brown, R., A role for methylation of the hMLH1 promoter in loss of hMLH1 expression and drug resistance in ovarian cancer, *Oncogene*, 18, 2335–2341, 1999.

Sun, W. H., Pabon, C., Alsayed, Y., Huang, P. P., Jandeska, S., Uddin, S., Platanias, L. C. and Rosen, S. T., Interferon-alpha resistance in a cutaneous T-cell lymphoma cell line is associated with lack of STAT1 expression, *Blood*, 91, 570–576, 1998.

Suzuki, H., Gabrielson, E., Chen, W., Anbazhagan, R., van Engeland, M., Weijenberg, M. P., Herman, J. G. and Baylin, S. B., A genomic screen for genes upregulated by demethylation and histone deacetylase inhibition in human colorectal cancer, *Nat. Genet.*, 31, 141–149, 2002.

Teitz, T., Wei, T., Valentine, M. B., Vanin, E. F., Grenet, J., Valentine, V. A., Behm, F. G., Look, A. T., Lahti, J. M. and Kidd, V. J., Caspase 8 is deleted or silenced preferentially in childhood neuroblastomas with amplification of MYCN, *Nat. Med.*, 6, 529–535, 2000.

Thomas, T. J. and Messner, R. P., Effects of lupus-inducing drugs on the B to Z transition of synthetic DNA, *Arthritis Rheum.,* 29, 638–645, 1986.

Turner, B. M., Cellular memory and the histone code, *Cell,* 111, 285–291, 2002.

Velicescu, M., Weisenberger, D. J., Gonzales, F. A., Tsai, Y. C., Nguyen, C. T. and Jones, P. A., Cell division is required for de novo methylation of CpG islands in bladder cancer cells, *Cancer Res.,* 62, 2378–2384, 2002.

Vesely, J., Cihak, A. and Sorm, F., Characteristics of mouse leukemic cells resistant to 5-azacytidine and 5-aza-2′-deoxycytidine, *Cancer Res.,* 28, 1995–2000, 1968.

Villar-Garea, A., Fraga, M. F., Espada, J. and Esteller, M., Procaine is a DNA-demethylating agent with growth-inhibitory effects in human cancer cells, *Cancer Res.,* 63, 4984–4989, 2003.

Weber, J., Salgaller, M., Samid, D., Johnson, B., Herlyn, M., Lassam, N., Treisman, J. and Rosenberg, S. A., Expression of the MAGE-1 tumor antigen is up-regulated by the demethylating agent 5-aza-2′-deoxycytidine, *Cancer Res.,* 54, 1766–1771, 1994.

Wilson, V. L., Jones, P. A. and Momparler, R. L., Inhibition of DNA methylation in L1210 leukemic cells by 5-aza-2′-deoxycytidine as a possible mechanism of chemotherapeutic action, *Cancer Res.,* 43, 3493–3496, 1983.

Wolf, S. F. and Migeon, B. R., Studies of X chromosome DNA methylation in normal human cells, *Nature,* 295, 667–671, 1982.

Wong, L. H., Krauer, K. G., Hatzinisiriou, I., Estcourt, M. J., Hersey, P., Tam, N. D., Edmondson, S., Devenish, R. J. and Ralph, S. J., Interferon-resistant human melanoma cells are deficient in ISGF3 components, STAT1, STAT2, and p48-ISGF3gamma, *J. Biol. Chem.,* 272, 28779–28785, 1997.

Yang, Q., Shan, L., Yoshimura, G., Nakamura, M., Nakamura, Y., Suzuma, T., Umemura, T., Mori, I., Sakurai, T. and Kakudo, K., 5-Aza-2′-deoxycytidine induces retinoic acid receptor beta 2 demethylation, cell cycle arrest and growth inhibition in breast carcinoma cells, *Anticancer Res.,* 22, 2753–2756, 2002.

Yogelzang, N. J., Herndon, J. E., 2nd, Cirrincione, C., Harmon, D. C., Antman, K. H., Corson, J. M., Suzuki, Y., Citron, M. L. and Green, M. R., Dihydro-5-azacytidine in malignant mesothelioma. A phase II trial demonstrating activity accompanied by cardiac toxicity. Cancer and Leukemia Group B, *Cancer,* 79, 2237–2242, 1997.

Zheng, T. S., Hunot, S., Kuida, K., Momoi, T., Srinivasan, A., Nicholson, D. W., Lazebnik, Y. and Flavell, R. A., Deficiency in caspase-9 or caspase-3 induces caspase activation, *Nat. Med.,* 6, 1241–1247, 2000.

Zhou, L., Cheng, X., Connolly, B. A., Dickman, M. J., Hurd, P. J. and Hornby, D. P., Zebularine: a novel DNA methylation inhibitor that forms a covalent complex with DNA methyltransferases, *J. Mol. Biol.,* 321, 591–599, 2002.

Zingg, J. M., Shen, J. C., Yang, A. S., Rapoport, H. and Jones, P. A., Methylation inhibitors can increase the rate of cytosine deamination by (cytosine-5)-DNA methyltransferase, *Nucleic Acids Res.,* 24, 3267–3275, 1996.

13 DNA Demethylating Agents: Preclinical Evaluation as Anticancer Agents

Robert Brown

CONTENTS

ACKNOWLEDGMENTS

R.B. acknowledges support from Cancer Research UK.

INTRODUCTION

All human cancers appear to show loss of the normal control of DNA methylation. Analysis of DNA methylation at a genome-wide level has demonstrated that tumors often exhibit overall decreased levels of DNA methylation [1]. In contrast, individual CpG islands often exhibit increased methylation in cancer. Indeed, methods that enable large-scale analysis of CpG islands throughout the genome indicate that cancers probably exhibit aberrant increased methylation of hundreds, or even thousands, of CpG islands within a single tumor [2,3]. The prevalence of aberrant DNA

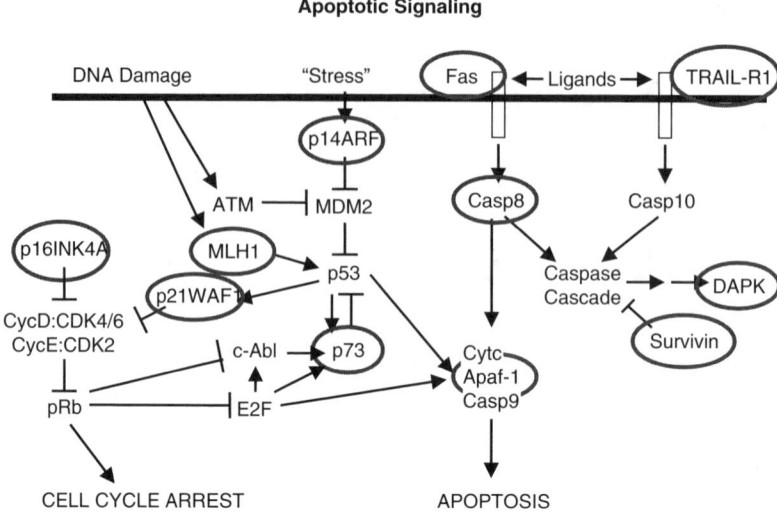

FIGURE 13.1 Apoptotic and cell cycle control pathways in cancer. Circled genes are frequently methylated and the gene epigenetically silenced in cancer.

methylation in cancer makes it a good target to consider for the development of novel anticancer agents. Many genes involved in apoptosis signaling pathways can frequently become hypermethylated and transcriptionally silenced in tumors (Figure 13.1). Reversal of gene methylation and epigenetic silencing, therefore, has the potential either to cause suppression of tumor growth or to increase the ability of tumor cells to undergo drug-induced apoptosis. In this chapter we discuss issues surrounding the development of DNA demethylating agents as potential anticancer agents and the types of approaches currently being used to evaluate them preclinically. These studies show that there is a need for robust measures of biological activity developed in the preclinical setting that can be used to evaluate clinical activity and that laboratory models be used to guide rational combinations of demethylating agents with other treatments.

DNA METHYLTRANSFERASES AS TARGETS

Key targets for potential DNA demethylating agents are DNA methyltransferases (DNMTs), enzymes that catalyze the addition of a methyl group to cytosine residues at CpG dinucleotides [4]. To date, three members of the DNMT family have been described in mammalian cells. DNMT1 is believed to function primarily to maintain the DNA methylation pattern following the synthesis of new DNA during cell division, as it exhibits much higher activity on hemimethylated DNA than on unmethylated DNA [5]. DNMT3a and 3b show no preference for hemimethylated DNA and, based on inactivation of the DNMT3a and 3b genes in mice, they are believed to function principally as *de novo* methyltransferases [6]. Genetic inactivation of DNMTs in the germ line of mice indicates that inactivation is developmentally lethal. However, this is not a major concern for the treatment of a somatic disease such as cancer with DNMT inhibitors. Simultaneous inactivation of DNMT1 and DNMT3b in a colon tumor cell line slows the growth of the cells and induces dramatic reversion of global methylation [7]. These marked changes result in demethylation of repeated sequences, loss of insulin-like growth factor II (IGF2) imprinting, abrogation of silencing of the tumor suppressor gene p16^{INK4a}, and growth suppression. Such genetic inactivation studies are important proof-of-principle experiments showing the central role of these genes in maintaining a hypermethylated phenotype in tumor cells and validating inhibition of DNMTs as targets for inducing demethylating activity.

Reduction in DNMT levels has been suggested to cause increased chromosomal instability in somatic cells, which could influence tumor progression or be carcinogenic [8]. However, these characteristics are similar to many clinically useful DNA damaging anticancer agents, where therapy-associated tumors can occur in a small proportion of patients many years after treatment [9]. Patients find this an acceptable risk for the treatment of a life-threatening illness such as cancer. More immediate concerns regarding the clinical use of DNA demethylating agents are:

- Can a biologically active demethylating dose of drug be given safely with acceptable toxicity?
- What is the optimum schedule of drug?
- Are there drug interactions that might increase toxicity or sensitivity/resistance of tumors to existing therapies?
- Can patient groups be identified who may particularly benefit from demethylating agents?

Preclinical experimental models can help shed light on these important clinical issues.

TYPES OF DEMETHYLATING AGENTS

NUCLEOSIDE DNMT INHIBITORS

The two closely related drugs 5-azacytidine and 2'deoxy-5-azacytidine, also known as decitabine, have long been used experimentally to inhibit DNA methylation in tissue culture and have been shown to reactivate numerous methylation-silenced genes (see Chapter 10). These nucleoside DNMT inhibitors are phosphorylated to their nucleotide prior to being incorporated into DNA. Once incorporated into DNA they complex and inactivate all three forms of DNA methyltransferases. Nucleoside DNMT inhibitors have been reported to have antitumor activity, especially against hematological malignancies. Like many other novel therapeutics currently being developed, demethylating agents (that do not have unwanted additional activities) are hoped not to function in a nonspecific manner and, thus, have fewer side effects than conventional chemotherapy. Rather, they have been proposed to reverse repression of tumor suppressor genes and cell cycle genes aberrantly methylated in tumor cells leading to inhibition of tumor growth. An important consequence of this is that, unlike conventional cytotoxic agents, it may be best to use such drugs at concentrations lower than the maximum tolerated dose. For instance, there is an optimal concentration at which analogues of 5-azacytosine induce cellular differentiation; higher concentrations produce less differentiation and more cytotoxicity [10]. Thus in the case of decitabine, although its use at high doses may induce direct toxicity effects due to its incorporation into DNA, prolonged low dose schedules [11] or combinations with other drugs [12] may be more biologically effective in inhibiting DNMT activity with less toxicity. For example, it has been demonstrated that, in mouse xenograft models, treatment with relatively low doses of decitabine can result in reactivation of a methylation-silenced gene, *MLH1* [12]. The MLH1 protein, part of the human DNA mismatch repair system, has been shown to be important in determining sensitivity to a number of chemotherapeutic agents and, indeed, the treated xenografts exhibited clear increases in sensitivity to drugs such as carboplatin, temozolomide, and epirubicin.

Decitabine exhibits poor oral bioavailability and, consequently, is administered intravenously. However, an oral DNMT inhibitor, zebularine [1-(beta-D-ribofuranosyl)-dihydropyrimidin-2-one], has recently been identified [13]. *In vitro* experiments have shown that synthetic oligonucleotides containing zebularine form tight complexes with bacterial methyltransferases, leading to a potent inhibition of DNA methylation [14]. The use of an oral DNA methyltransferase inhibitor may become important if prolonged low dose administration and constant enzyme inhibition is necessary to produce a biological effect.

One of the limitations of the nucleoside analogue DNMT inhibitors in clinical trials has been treatment-associated side effects, such as neutropenia and thrombocytopenia [15]. It is believed that these cytotoxic effects are due to effects of the drug after incorporation into DNA that are not related to DNMT inhibition. Indeed, the types of toxicities observed in mice and humans at these maximum tolerated doses are consistent with DNA damaging activity rather than DNA demethylation activity.

Nonnucleoside DNMT Inhibitors

The toxicity of nucleoside DNMT inhibitors has stimulated a search for DNMT inhibitors that are not incorporated into DNA. The drug procainamide (approved by the U.S. Food and Drug Administration for the treatment of cardiac arrythmias) has been suggested to be a nonnucleoside inhibitor of DNMT [16]. In human prostate cell lines and xenograft tumors, procainamide can reverse GSTP1 CpG island hypermethylation and restore GSTP1 expression.

Another approach being used to inhibit DNA methylation is the use of antisense oligonucleotides. Antisense oligonucleotides directed against the DNMT1 mRNA have been shown to reduce DNMT1 protein levels and induce demethylation and expression of the p16 tumor suppressor gene in human tumor cells [17] and also inhibits tumor growth in mouse models [18]. This DNMT1 antisense molecule has also been used in phase I clinical trials and has shown some antitumor activity (for further details see http://www.methylgene.com/html).

PHARMACODYNAMIC MEASURES OF DEMETHYLATING AGENTS

The maximum biologically effective dose of a DNA methylating agent may not equate to the maximum tolerated dose. Early clinical trials that address safety of a drug examine maximum tolerated dose, but an important element of clinical trials involving methylation inhibitors should be the use of molecular endpoints to monitor changes in DNA methylation. Clearly the most satisfactory approach would be to monitor levels of methylation of important tumor suppressor genes in tumor DNA. However, such an approach would be possible only in a limited number of tumor types, such as hematological malignancies, where repeated sampling of the tumor was feasible (see Chapter 12). Another approach is to determine total genomic levels of 5′methylcytosine in a surrogate tissue, such as peripheral blood mononuclear cells. It has previously been shown, using an HPLC-based method, that reductions in total genomic levels of 5′methylcytosine in peripheral blood mononuclear cells of xenograft bearing mice closely mirrors reduced methylation of the *MLH1* promoter following decitabine treatment [12]. Similarly, genes methylated in normal tissue, such as imprinted genes [19], can also be assessed for reductions in methylation in surrogate tissues. Fetal hemoglobin is repressed in adults by DNA methylation and levels are increased by treatment with inhibitors of DNA methylation [20] and, thus, represent another potential molecular endpoint to monitor the effect of DNA methylation inhibitors. MAGE1a is biallelically methylated in most normal cells and contains a CpG island in the promoter region. Biallelic methylation of MAGE1a makes it easier to monitor changes in methylation using techniques such as methylation-specific PCR, COBRA, or sequencing of bisulfite-modified DNA in surrogate tissues such as peripheral blood mononuclear cells or buccal smears.

The end result of gene activation with inhibitors of DNA methylation, such as decitabine, can be the induction of differentiation of undifferentiated cells and activation of tumor suppressor genes [21,22]. There is an optimal concentration at which analogues of 5-azacytosine induce cellular differentiation; higher concentrations produce less differentiation and more cytotoxicity [10]. Morphological and functional differentiation is induced in primary suspension cultures of human leukemia cells following repeated exposure to decitabine [23]. Cellular differentiation after treatment with decitabine has been observed in many cell systems, although the induction of differentiation may be dependent on the cell line studied.

High-density oligonucleotide gene expression microarrays have been used to examine the effects of decitabine treatment on human fibroblast cells (LD419) and a human bladder tumor cell line (T24) [24]. Data obtained 8 days after recovery from decitabine treatment showed that more genes were induced in tumorigenic cells (61 genes induced; ≥4-fold) than nontumorigenic cells (34 genes induced; ≥4-fold). Approximately 60% of induced genes did not have CpG islands within their 5′ regions, suggesting that some genes activated by decitabine may not result from the direct inhibition of promoter methylation. Interestingly, a high percentage of genes activated in both cell types belonged to the interferon signaling pathway.

Many reports have been published on the activation of tumor suppressor genes by decitabine (see also Chapter 10). Herman et al., for example, reported that decitabine activated the tumor suppressor gene VHL in human renal carcinoma cell lines [25]. Decitabine can activate the expression of p16/CDKN2 tumor suppressor gene in lung cancer cell lines and in some lung tumor surgical specimens [26,27]. The p16/CDKN2 tumor suppressor gene inhibits the enzyme activity of cyclin-dependent kinases that mediate the phosphorylation of the retinoblastoma (Rb) gene product. The hypophosphorylation of Rb is thought to prevent the entry of tumor cells into the S-phase of the cell cycle.

Decitabine can activate the expression of E-cadherin in primary tumors of breast and prostate [28,29]. E-cadherin is a cell-adhesion molecule that suppresses tumor cell invasion and metastasis in experimental tumor models. Estrogen receptor (ER) can become methylated in breast tumors and reactivated by decitabine in ER-negative human breast carcinoma cell lines [30].

What can be concluded from these many studies on the effects of DNA demethylation on gene expression in tumor cells is that many can be altered in expression by both direct and indirect mechanisms. Reexpression of genes in tumors will have different potential effects. For instance, reexpression of some genes will be pro-apoptotic while others will be antiapoptotic. Measures of 5-methylcytosine levels and methylation of CpG islands will give pharmacodynamic measures that demethylation is being induced; however, this will not identify the key targets that will inhibit growth of a given tumor type. Indeed the effects of demethylating agents may vary between tumor types, and between patients, depending on the profile of genes aberrantly methylated in a given tumor.

DRUG COMBINATIONS

COMBINATIONS WITH EXISTING THERAPIES

Re-silencing of gene expression does recur over time in cells after treatment with DNMT inhibitors such as decitabine [12]. Therefore, there may be only a specific window of time when tumor cells will die due to epigenetic reversal of silencing of tumor suppressor genes and subsequent apoptosis. On the other hand, this window of demethylation can be utilized for appropriate scheduling of a cytotoxic or other treatments whose antitumor effects are being inhibited due to gene methylation.

Synergistic cytotoxicity of decitabine and a variety of clinically important cytotoxic agents has been reported *in vitro* and *in vivo*, including cisplatin, carboplatin, epirubicin, cyclophosphamide, and temozolomide [12,31]. The combination of decitabine and retinoic acid produced an additive antileukemic effect in HL-60 cells, which correlated with additive reduction of *c-myc* expression and increase in the differentiation of the leukemia cells [32].

The combination of decitabine and cisplatin showed a synergistic cytotoxic interaction in many human tumor cell lines. Although a possible underlying mechanism originally suggested is the increased binding of cisplatin to decitabine-substituted DNA that is independent of DNA hypomethylation [33], more recent studies have focused on the effects of decitabine in reactivating drug sensitivity genes [34]. Loss of DNA mismatch repair (MMR) because of hypermethylation of the *hMLH1* gene promoter occurs at a high frequency in a number of human tumors. A role for loss of MMR in resistance to a number of clinically important anticancer drugs has been shown [12]. Decitabine was used *in vivo* to sensitize MMR-deficient, drug-resistant ovarian (A2780/cp70) and colon (SW48) tumor xenografts that are MLH1 negative because of gene promoter hypermethylation [12]. Treatment of tumor-bearing mice with the demethylating agent decitabine at a nontoxic dose induces MLH1 expression. Reexpression of MLH1 was associated with a decrease in *hMLH1* gene promoter methylation. Decitabine treatment alone had no effect on the growth rate of the tumors. However, decitabine treatment sensitized the xenografts to cis-platin, carboplatin, temozolomide, and epirubicin. Sensitization was comparable with that obtained by reintroduction of the *hMLH1* gene by chromosome 3 transfer. Decitabine treatment did not sensitize xenografts of HCT116, which lacks MMR because of *hMLH1* mutation.

The human *multidrug resistance gene 1 (MDR1)* encodes P-glycoprotein (Pgp), a transmembrane protein that acts as a drug efflux pump, reducing intracellular levels of certain anticancer drugs, and thus reducing their effectiveness. Increased *MDR1* expression has been shown to be a clinically relevant mechanism for leukemia, although its relevance for solid tumors remains controversial [35]. Increased tran-scription of the MDR1 gene in chronic lymphocytic leukemia and bladder cancer following chemotherapy has been shown to be associated with decreased methylation [36,37]. This would argue that treatment of sensitive tumors with a demethylating agent could lead to resistance to chemotherapy by increased expression of *MDR1*. Indeed, increased resistance of tumor cells after treatment with azacytidine analogues to drugs that are substrates of Pgp has been observed [37]. However, increased sensitization and no effect has also been reported to be induced by DNMT inhibitors for MDR drugs in different tumor models [12,38,39]. This again emphasizes the possibility that these agents will have different effects depending on the pattern of genes methylated in a given tumor and argues that patient stratification, depending on their methylation status, may be necessary in clinical trials of demethylating agents.

COMBINATIONS WITH HISTONE DEACETYLASE INHIBITORS

Initially, it was suggested that DNA methylation inhibited transcription factor bind-ing leading to suppression of gene transcription. Indeed, several important transcrip-tion factors have been shown to be sensitive to methylation of CpGs within their recognition sites [40]. In recent years though, a more generally applicable mecha-nism, by which DNA methylation can lead to transcriptional repression, has begun to be elucidated [41,42]. DNA methylation leads to the binding of a family of proteins known as methyl-binding domain (MBD) proteins. The members of this protein family all share a common methyl-binding domain that enables them to bind spe-cifically to DNA that contains methylated CpG sites. At least three of the five known members of this family (MeCP2, MBD2, MBD3) have been shown to be associated with large protein complexes containing histone deacetylase (HDAC1 and HDAC2) and chromatin remodeling activities (sin3a and mi-2). Thus, DNA methylation has a direct influence on both histone acetylation and higher-order chromatin structure. The action of these histone deacetylases (HDACs) is to restore a positive charge to lysine residues in the amino tail of histones. This, together with other chromatin remodeling activities, is then thought to result in the production of compacted chromatin that is refractory to transcription.

Although these observations argue that DNA methylation is a key signal leading to histone modifications, chromatin remodeling and gene silencing, this signaling can also operate in the opposite direction. For example, disruption of histone methylation in *Neurospora crassa* results in the elimination of DNA methylation [43]. Disruption of histone methylation can be brought about by mutation of the *dim-5* gene, which encodes a protein homologous to the chromatin-associated protein Suv39h found in mammalian cells. Similarly, increased histone acetylation in cells treated with HDAC inhibitors can also lead to demethylation of DNA [44]. Another key connection between DNA methylation and histone acetylation has been established by the finding that DNMT1 associates with HDAC1 and is part of a complex that also contains the RB tumor suppressor gene product and the transcription factor E2F. DNMT1 cooperates with other cellular factors to repress transcription from promoters containing E2F-binding sites [45,46]. It has been suggested that loss of function of RB, a frequent event in several types of tumors, might result in improper regulation of this complex, which results in mislocalization of DNMT1 and the production of aberrant methylation patterns. An important implication of these results is that if changes in histones can influence DNA methylation, then it is possible that the methylation of genes observed in tumors is a result of transcriptional silencing owing to histone modification or chromatin remodeling, rather than of DNA methyltransferase activity per se.

Taken together, these studies demonstrate the emerging concept that crosstalk between these different mechanisms of epigenetic regulation (DNA methylation, histone methylation and acetylation, chromatin remodeling) is essential for appropriate control of gene transcription. The different epigenetic layers engaging in this complex crosstalk means that drug development strategies must consider what are the crucial targets for an effective epigenetic drug and how these drugs might be used in combination. For instance, to date DNA methyltransferase inhibitors, such as decitabine, appear to be the most active compounds for inducing reexpression of epigenetically silenced genes in tumor cell models. However, HDAC inhibitors can increase levels of gene expression and have been shown to work together with DNA methyltransferase inhibitors to induce gene reexpression [47]. Cameron et al. demonstrated that combining treatment with decitabine and treatment with an inhibitor of histone deacetylase, trichostatin A, caused a synergistic reversal of transcriptional silencing of the *MLH1* and *TIMP3* genes in the colorectal cancer cell line RKO [47]. We have also been able to show that novel hydroxamic acid HDAC inhibitors can synergize with decitabine in inducing reexpression of MLH1 in drug-resistant tumor models (ovarian, colon, and lung), and this leads to increased sensitivity to chemotherapeutic drugs (unpublished).

Families of HDACs and histone acetyl transferases (HATs) determine the acetylation of histones [48,49]. At least five groups of proteins with HAT activity and three classes of HDACs play a role in modeling the structure of chromatin. In addition to histone acetylation, these enzymes regulate gene expression by acetylating or deacetylating transcription factors such as p53, GATA-1, and TFIIE. Expression of different

HATs and HDACs or different patterns of histone acetylation at gene promoters in cancer has not been extensively studied, although certain cancer-prone syndromes and tumors have been associated with mutations in a HAT or aberrant recruitment of HDACs [50,51]. HDAC inhibitors have been shown to inhibit proliferation of tumor cell lines *in vitro* and in some cases induce apoptosis [50,51]. Several structural classes of HDAC inhibitors have been examined:

- Short-chain fatty acids (e.g., butyrates)
- Hydroxamic acids (e.g., trichostatin A, suberoylanilide hydroxamic acid, oxamflatin)
- Cyclic tetrapeptides (e.g., trapoxin A, apicidin)
- Benzamides (e.g., MS-275).

The structure of the catalytic core of HDACs has been determined by radiograph crystallography [52]. Residues that form the active site are conserved across all HDACs, and hydroxamic acid HDAC inhibitors have been shown to complex through zinc at a zinc-binding site in a tubular pocket at the active catalytic site.

HDAC inhibitors, in particular hydroxamic acid compounds, inhibit purified HDACs at nanomolar concentrations and induce growth arrest, differentiation, and apoptosis in transformed cells. Using acetylated histone-specific antibodies, tumor cells or normal cells from animals treated with HDAC inhibitors show accumulation of acetylated histones. The accumulation of acetylated histones in peripheral mononuclear cells has potential as a pharmacodynamic measure of biological activity of HDAC inhibitors in clinical trials. In addition, specific genes, such as the cyclin-dependent kinase inhibitor p21/WAF1, have been shown to be upregulated by HDAC inhibitors and measurement of the levels of these genes could also be of value as pharmacodynamic endpoints in clinical trials [48,50,51].

TUMOR PROGNOSIS AND STRATIFICATION

Another potential use of DNA methylation is in the classification of tumors depending on their methylation status. Such classification might be of use in determining patient prognosis or potential response to therapy. Indeed, a number of DNA methylation studies have already identified links between methylation and patient outcome. For example, a study of methylation of the DNA repair gene *MGMT* identified a clear link between methylation of the *MGMT* promoter and increased overall and disease-free survival [53], probably due to increased responsiveness to alkylating agents in MGMT-deficient tumors. Similarly, Tang et al. [54] determined that increased methylation of the *DAP kinase* gene was strongly associated with decreased survival in patients with non-small-cell lung carcinoma and found that *DAP kinase* methylation was probably the strongest independent prognostic factor in these patients. In addition, the development of methods for the large-scale analysis of CpG island methylation referred to above raises the prospect of dramatically increasing our ability to classify tumors based on DNA methylation. These methods have already been able to identify differences between methylation patterns in different tumors [3], correlations between methylation and tumor grade [2], and shown their potential to identify which patients may benefit from chemotherapy [55].

The prevalence of aberrant CpG island methylation and associated epigenetic silencing of genes in tumors makes it an attractive target for novel anticancer therapies. Several small molecules are now entering early clinical trials that can reverse demethylation or alter chromatin modification. Given the crosstalk between epigenetic pathways, combinations of these drugs could be more effective than individual drugs. Combinations of epigenetic drugs with conventional anticancer therapies could also be important given their transient effects on gene reexpression. Pharmacodynamic measures of DNA methylation and histone acetylation, as well as subsequent effects on gene expression, might help to drive the appropriate scheduling of these combinations. Epigenomic analysis of patterns of global DNA methylation and chromatin positioning will help to identify those patients who will particularly benefit from DNA demethylating agents.

REFERENCES

1. Costello, J.F. and Plass, C., Methylation matters. *J. Med. Genet.*, 38, 285–303, 2001.
2. Yan, P.S., Perry, M.R., Laux, D.E., Asare, A.L., Caldwell, C.W. and Huang, TH-M., CpG island arrays: an applicaion towards deciphering epigenetic signatures of breast cancer. *Clin. Cancer Res.*, 6, 1432–1438, 2000.
3. Costello, J.F., Fruhwald, M.C., Smiraglia, D.J., Rush, L.J., Robertson, G.P., Gao, X., Wright, F.A., Feramisco, J.D., Peltomaki, P., Lang, J.C. et al., Aberrant CpG-island methylation has non-random and tumour-type-specific patterns, *Nature Genetics*, 24, 132–138, 2000.
4. Hendrich, B. and Bird, A., Mammalian methyltransferases and methyl-CpG-binding domains: proteins involved in DNA methylation, *Curr. Top. Microbiol. Immunol.*, 249, 55–74, 2000.
5. Bestor, T.H. and Verdine, G.L., DNA methyltransferases, *Curr. Opin. Cell. Biol.*, 6, 380–389, 1994.
6. Okano, M., Bell, D.W., Haber, D.A. and Li, E., DNA methyltransferases Dnmt3a and Dnmt3b are essential for de novo methylation and mammalian development, *Cell*, 99, 247–257, 1999.
7. Rhee, I., Bachman, K.E., Park, B.H., Jair, K.W., Yen, R.W., Schuebel, K.E., Cui, H., Feinberg, A.P., Lengauer, C., Kinzler, K.W. et al., DNMT1 and DNMT3b cooperate to silence genes in human cancer cells, *Nature*, 416, 552–556, 2002.
8. Gaudet, F., Hodgson, J.G., Ede, A., Jackson-Grusby, L., Dausman, J., Gray, J.W., Leonhardt, H. and Jaenisch, R. Induction of tumors in mice by genomic hypomethylation, *Science*, 300:489–492, 2003.
9. Smith, S.M., Le Beau, M.M., Huo, D., Karrison, T., Sobecks, R.M., Anastasi, J., Vardiman, J.W., Rowley, J.D. and Larson, R.A., Clinical-cytogenetic associations in 306 patients with therapy-related myelodysplasia and myeloid leukemia: the University of Chicago series, *Blood*, 102, 43–52, 2003.
10. Taylor, S.M. and Jones, P.A., Multiple new phenotypes induced in 10T2 and 3T3 cells treatefd with 5-azacytidine, *Cell*, 17, 771–779, 1979.

11. Lubbert, M., Wijermans, P., Kunzmann, R., Verhoef, G., Bosly, A., Ravoet, C., Andre, M. and Ferrant, A., Cytogenetic responses in high-risk myelodysplastic syndrome following low-dose treatment with the DNA methylation inhibitor 5-aza-2'-deoxycytidine, *Br. J. Haematology*, 114, 349–357, 2001.

12. Plumb, J.A., Strathdee, G., Sludden, J., Kaye, S.B. and Brown, R., Reversal of drug resistance in human tumour xenografts by 2'deoxy-5-azacytidine-induced demethylation of the *hMLH1* gene promoter, *Cancer Res.*, 60, 6039–6044, 2000.

13. Cheng, J.C., Matsen, C.B., Gonzales, F.A., Ye, W., Greer, S., Marquez, V.E., Jones, P.A. and Selker, E.U., Inhibition of DNA methylation and reactivation of silenced genes by zebularine, *J. Natl. Cancer Inst.*, 95, 399–409, 2003.

14. Zhou, L., Cheng, X., Connolly, B.A., Dickman, M.J., Hurd, P.J. and Hornby, D.P., Zebularine: a novel DNA methylation inhibitor that forms a covalent complex with DNA methyltransferases, *J. Mol. Biol.*, 321, 591–599, 2002.

15. Momparler, R.L., Cote, S. and Eliopoulos, N., Pharmacological approach for optimisation of the dose schedule of 5-aza-2'-deoxycitidine (Decitabine) for the therapy of leukaemia, *Leukemia*, 11, 175–180, 1997.

16. Lin, X., Asgari, K., Putzi, M.J., Gage, W.R., Yu, X., Cornblatt, B.S., Kuma, A., Piantadosi, S., DeWeese, T.L., De Marzo, A.M. et al., Reversal of GSTP1 CpG island hypermethylation and reactivation of pi-class glutathione S-transferase (GSTP1) expression in human prostate cancer cells by treatment with procainamide, *Cancer Res.*, 61, 8611–8616, 2001.

17. Fournel, M., Sapieha, P., Beaulieu, N., Besterman, J.M., Macleod, A.R., Downregulation of human DNA-(cytosine-5) methyltransferase induces cell cycle regulators p16(ink4A) and p21(WAF/Cip1) by distinct mechanisms, *J. Biol. Chem,*. 274, 24250–24256, 1999.

18. Ramchandani, S., MacLeod, A.R., Pinard, M., von Hofe, E. and Szyf, M., Inhibition of tumorigenesis by a cytosine-DNA, methyltransferase, antisense oligodeoxynucleotide., *Proc. Natl. Acad. Sci. USA*, 94, 684–689, 1997.

19. Bartolomei, M.S. and Tilghman, S.M,, Genomic imprinting in mammals. *Ann. Rev. Genetics*, 31, 493–525, 1997.

20. DeSimone, J., Koshy, M., Dorn, L., Lavelle, D., Bressler, L., Molokie, R. and Talischy, N., Maintenance of elevated fetal hemoglobin levels by decitabine during dose interval treatment of sickle cell anemia, *Blood*, 99, 3905–3908, 2002.

21. Momparler, R.L., Bouchard, J. and Samson, J., Induction of differentiation and inhibition of DNA methylation in HL-60 myeloid leukemic cells by 5-aza-2'-deoxycytidine. *Leuk. Res.*, 9, 1361–1366, 1985.

22. Brown, R. and Strathdee, G., Epigenomics and epigenetic therapy of cancer, *Trends Mol. Med.*, 8, S43–S48, 2002.

23. Pinto, A., Attadia, V., Fusco, A., Ferrara, F., Spada, O.A. and Di Fiore, P.P., 5-Aza-2'-deoxycytidine induces terminal differentiation of leukemic blasts from patients with acute myeloid leukemias, *Blood*, 64, 922–929, 1984.

24. Liang, G., Gonzales, F.A., Jones, P.A., Orntoft, T.F. and Thykjaer, T., Analysis of gene induction in human fibroblasts and bladder cancer cells exposed to the methylation inhibitor 5-aza-2'-deoxycytidine, *Cancer Res.*, 62, 961–966, 2002.

25. Herman, J.G., Latif, F., Weng, Y., Lerman, M.I., Zbar, B., Liu, S., Samid, D., Duan, D.S., Gnarra, J.R., Linehan, W.M. et al., Silencing of the VHL tumor-suppressor gene by DNA methylation in renal carcinoma, *Proc. Natl. Acad. Sci. USA*, 91, 9700–9704, 1994.

26. Otterson, G.A., Khleif, S.N., Chen, W., Coxon, A.B. and Kaye, F.J., CDKN2 gene silencing in lung cancer by DNA hypermethylation and kinetics of p16INK4 protein induction by 5-aza 2'deoxycytidine, *Oncogene*, 11, 1211–1216, 1995.

27. Merlo, A., Herman, J.G., Mao, L., Lee, D.J., Gabrielson, E., Burger, P.C., Baylin, S.B. and Sidransky, D., 5' CpG island methylation is associated with transcriptional silencing of the tumour suppressor p16/CDKN2/MTS1 in human cancers, *Nat. Med.*, 1, 686–692, 1995.

28. Graff, J.R., Gabrielson, E., Fujii, H., Baylin, S.B. and Herman, J.G., Methylation patterns of the E-cadherin 5' CpG island are unstable and reflect the dynamic, heterogeneous loss of E-cadherin expression during metastatic progression, *J. Biol. Chem.*, 275, 2727–2732, 2000.

29. Nass, S.J., Herman, J.G., Gabrielson, E., Iversen, P.W., Parl, F.F., Davidson, N.E. and Graff, J.R., Aberrant methylation of the estrogen receptor and E-cadherin 5' CpG islands increases with malignant progression in human breast cancer, *Cancer Res.*, 60, 4346–4348, 2000.

30. Yang, X., Phillips, D.L., Ferguson, A.T., Nelson, W.G., Herman, J.G. and Davidson, N.E., Synergistic activation of functional estrogen receptor (ER)-alpha by DNA methyltransferase and histone deacetylase inhibition in human ER-alpha-negative breast cancer cells, *Cancer Res.*, 61, 7025–7029, 2001.

31. Frost, P., Abbruzzese, J.L., Hunt, B., Lee, D, and Ellis, M., Synergistic cytotoxicity using 2'-deoxy-5-azacytidineand cisplatin or 4-hydroperoxycyclophosphamide with human tumor cells, *Cancer Res.*, 50, 4572–4577, 1990.

32. Momparler, R.L., Dore, B.T. and Momparler, L.F., Effect of 5-aza-2'-deoxycytidine and retinoic acid on differentiation and c-myc expression in HL-60 myeloid leukemic cells, *Cancer Lett.*, 54, 21–28, 1990.

33. Ellerhorst, J.A., Frost, P., Abbruzzese, J.L., Newman, R.A. and Chernajovsky, Y., 2'-deoxy-5-azacytidine increases binding of cisplatin to DNA by a mechanism independant of DNA hypomethylation, *Brit. J. Cancer*, 67, 209–215, 1993.

34. Strathdee, G., MacKean, M., Illand, M. and Brown, R., A role for methylation of the *hMLH1* promoter in loss of hMLH1 expression and drug resistance in ovarian cancer *Oncogene*, 18, 2335–2341, 1999.

35. Links, M. and Brown, R., Clinical relevance of the molecular mechanisms or resistance to anti-cancer drugs, *Exp. Rev. Mol. Med.*, 1999, http://www3.cbcu.cam.ac.uk/HyperNews/get/txt001rbg.html.

36. Tada, Y., Wada, M., Kuroiwa, K., Kinugawa, N., Harada, T., Nagayama, J., Nakagawa, M., Naito, S. and Kuwano, M., MDR1 gene overexpression and altered degree of methylation at the promoter region in bladder cancer during chemotherapeutic treatment, *Clin. Cancer Res.*, 6, 4618–4627, 2000.

37. Kantharidis, P., El-Osta, A., deSilva, M., Wall, D.M., Hu, X.F., Slater, A., Nadalin, G., Parkin, J.D. and Zalcberg, J.R., Altered methylation of the human MDR1 promoter is associated with acquired multidrug resistance, *Clin. Cancer Res.*, 3, 2025–2032, 1997.

38. Efferth, T., Futscher, B.W. and Osieka, R., 5-Azacytidine modulates the response of sensitive and multidrug-resistant K562 leukemic cells to cytostatic drugs, *Blood Cells Mol. Dis.*, 27, 637–648, 2001.

39. Ando, T., Nishimura, M. and Oka, Y., Decitabine (5-Aza-2'-deoxycytidine) decreased DNA methylation and expression of MDR-1 gene in K562/ADM cells, *Leukemia*, 14, 1915–1920, 2000.

40. Tate, P.H. and Bird, A., Effects of DNA methylation on DNA binding proteins and gene expression, *Curr. Op. Genet. Dev.*, 3, 226–231, 1993.

41. Bird, A. and Wolffe, A.P., Methylation-induced repression-belts, braces and chromatin. *Cell*, 99, 451–454, 1999.

42. Tyler, J.K. and Kadonaga, J.T., The "dark side" of chromatin remodelling: repressive effects on transcription, *Cell*, 99, 443–446, 1999.

43. Tamaru, H. and Selker, E.U., A histone H3 methyltransferase controls DNA methylation in *Neurospora crassa*, *Nature*, 414, 277–283, 2001.

44. Cervoni, N. and Szyf, M., Demethylase activity is directed by histone acetylation, *J. Biol. Chem.*, 276, 40778–40787, 2001.

45. Fuks, F., Burgers, W.A., Brehm, A., Hughes-Davies, L. and Kouzarides, T., DNA methyltransferase Dnmt1 associates with histone deacetylase activity, *Natl. Genet.*, 24, 88–91, 2001.

46. Robertson, K.D., Ait-Si-Ali, S., Yokochi, T., Wade, P.A., Jones, P.L. and Wolffe, A.P., DNMT1 forms a complex with Rb, E2F1 and HDAC1 and represses transcription from E2F-responsive promoters, *Natl. Genet.*, 25, 338–342, 2001.

47. Cameron, E.E., Bachman, K.E., Myohanen, S., Herman, J.G. and Baylin, S.B., Synergy of demethylation and histone deacetylase inhibition in the re-expression of genes silenced in cancer, *Nat. Genet.*, 21:103–107, 1999.

48. Marks, P.A., Richon, V.M., Breslow, R. and Rifkind, R.A., Histone deacetylase inhibitors as new cancer drugs, *Curr. Opin. Oncol.*, 13, 477–483, 2002.

49. Gray, G.G. and Ekstrom, T.J., The human histone deacetylase family, *Exp. Cell Res.*, 262, 75–83, 2001.

50. Plumb, J.A., Finn, P.W., Williams, R.J., Bandara, M.J., Romero, M.R., Watkins, C.J., La Thangue, N.B. and Brown, R., Pharmacodynamic response and inhibition of growth of human tumor xenografts by the novel histone deacetylase inhibitor PXD101, *Mol. Cancer Ther.*, 2, 721–728, 2003.

51. Marks, P.A., Richon, V.M. and Rifkind, R.A, Histone deacetylase inhibitors: inducers of differentiation or apoptosis of transformed cells, *J. Natl. Cancer Inst.*, 13, 477–483, 2000.

52. Finnin, M.S., Donigian, J.R., Cohen, A., Richon, V.M., Rifkind, R.A., Marks, P.A., Breslow, R. and Pavletich, N.P., Structure of a histone deacetylase homologue bound to the TSA and SAHA inhibitors, *Nature*, 401, 188–193, 1999.

53. Esteller, M., Garcia-Foncillas, J., Andion, E., Goodman, S.N., Hidalgo, O.F., Vanaclocha, V., Baylin, S. and Herman, J.G., Inactivation of the DNA repair gene MGMT and the clinical response of gliomas to alkylating agents, *New Eng. J. Med.*, 343, 1350–1354, 2000.

54. Tang, X., Khuri, F.R., Lee, J.J., Kemp, B.L., Liu, D., Hong, W.K., and Mao, L., Hypermethylation of the death-associated protein (DAP) kinase promoter and aggressiveness in stage I non-small call lung cancer, *J. Natl. Cancer Inst.*, 92, 1511–1516, 2000.

55. Wei, S.H., Chen, C.-M., Strathdee, G., Harnsomburana, J., Shyu, C.-R., Rahmatpanah, F., Shi, H., Ng, S.-W., Yan, P.S., Nephew, K.P. et al., Methylation microarray analysis of late-stage ovarian carcinomas distinguishes progression-free survival in patients and identifies candidate epigenetic markers, *Clin. Cancer Res.*, 8, 2246–2252, 2002.

14 DNA Demethylating Agents: Clinical Uses

Michael Lübbert, Jens Hasskarl, and Rainer Claus

CONTENTS

ACKNOWLEDGMENTS

We wish to thank Professor G. Schwartsmann and Professor M. S. Fernandez (Porto Alegre, Brazil) for kindly providing clinical follow-up data, Professor R. Mertelsmann for continued, helpful discussions, Mrs. B. Widmann for excellent secretarial assistance, and Dr. J. Pitako for major help with tabulation. M. L. is supported by Wilhelm-Sander Stiftung, Deutsche Forschungsgemeinschaft, and José Carreras Foundation. J. H. is supported by a grant from the University of Freiburg.

CLINICAL DEVELOPMENT OF DRUGS WITH DNA DEMETHYLATING ACTIVITY: FROM CYTOTOXIC TO EPIGENETIC TREATMENT

Pharmacologic reversal of epigenetic gene silencing by agents that interfere with DNA (cytosine) methylation provides a novel treatment approach that may be considered fundamentally different from the cytotoxic principle used in cancer treatment. Therefore, a better understanding of the mechanisms of action resulting in DNA demethylation *in vitro* and *in vivo* should aid in developing this novel treatment modality for different neoplasias and hemoglobinopathies. The two compounds that have advanced farthest in the clinical development for treatment of myelodysplastic syndrome (MDS) were originally synthesized and applied at high doses, with often severe toxicities, like other "classic" cytotoxic antimetabolites such as cytarabine. In 1964, Sorm and coworkers synthesized the ribonucleoside 5-azacytidine and its analogue, the deoxynucleoside 5-aza 2′-deoxycytidine (decitabine). Both differ from cytidine and 2′-deoxycytidine, respectively, by addition of a nitrogen residue at the fifth carbon position. Whereas the phosphorylation step to monophosphate is identical for both compounds, 5-azacytidine is predominantly incorporated into newly synthesized RNA (tRNA and mRNA). A minor proportion of 5-azacytidine is converted to 5-aza-2′-deoxycytidine, thus acting as its prodrug. In contrast, 5-aza-2′-deoxycytidine is exclusively incorporated into newly synthesized DNA and is thus the 5- to 10-fold more powerful DNA demethylating agent. Both drugs share the inactivation step of deamination by cytidine deaminase.

Both azanucleoside drugs showed a dose-dependent growth inhibitory effect on leukemia and tumor cell lines in tissue culture and, at high concentrations, cytotoxic effects that may be mediated by DNA strand breaks following the synthesis of alkali-labile DNA [1]. Both compounds were tested in rodent models for their antileukemic activity and were found to have more potent effects than cytarabine in this regard [2,3].

5-Azacytidine found rapid introduction into clinical trials: Hrodeck and Vesely first demonstrated antileukemic activity of this compound when administered intramuscularly to children with acute lymphoblastic leukemia (ALL) [4]. In parallel studies in adult patients with acute leukemia, the maximal tolerated dose (300 to 1125 mg/m^2 intravenously) was determined. Major hematologic toxicities were seen, as well as major nonhematologic toxicities, such as nausea and vomiting, diarrhea, and hepatic toxicity. Until the late 1970s, this drug had been used in at least 150 clinical trials in acute leukemia, almost always in combination with other drugs, and, except for one trial [5], not in first-line treatment of these diseases in large trials (see below). Decitabine followed a similar route of clinical development, being introduced into clinic by the Montreal Group; in 1985, Rivard, Momparler, and colleagues first used it in childhood leukemia [6]. The antileukemic effects at high doses as well as the toxicity profile were quite comparable to the experience with 5-azacytidine.

The clinical development and experience with azanucleosides up to the late 1970s were not much different from those of cytarabine. The fundamental discovery of the DNA demethylating activity and differentiation-inducing capacity of both drugs in tissue culture by Jones and Taylor opened a new avenue for development of both drugs at low-dose schedules, by exploring their biological, gene modulatory activities [7]. Constantinides et al., followed by Konieczny and Emerson, were the first to demonstrate induction of differentiation of embryonic fibroblasts to mature myocytes, adipocytes and chondrocytes [8,9]. Demethylation and transcriptional upregulation of MyoD, the master switch regulator of myocyte differentiation, was demonstrated briefly thereafter [10]. Building on this discovery, clinical investigators rapidly translated these results to clinical approaches of low-dose treatment. This included patients with hemoglobinopathies (with the rationale of reactivating fetal hemoglobin gamma gene which is developmentally silenced by methylation), preleukemic myelodysplastic syndromes, and acute myeloid leukemia (with the rationale of inducing myeloid differentiation). These trials showed a very different toxicity profile, with negligible nonhematologic toxicity, and less myelotoxicity than seen in the earlier phase of development. In the following, we will review the clinical results of high-dose and low-dose schedules of these azanucleosides in hematopoietic malignancy, their development for solid tumor patients, and patients with hemoglobinopathies. In addition, the present knowledge of *in vivo* activity of both demethylating drugs in translational studies is addressed. Finally, the concerns that azanucleosides, by being incorporated into DNA, may have a mutagenic and tumorigenic long-term adverse effect are discussed.

In addition to these azanucleosides, several other compounds have shown *in vitro* demethylating activity, some of which are advancing to clinical trials (Table 14.1). A novel compound, zebularine, has shown a strong inhibition of DNA methylation [11]. Several other recent studies have demonstrated, at least *in vitro*, demethylating activity of compounds not previously known for this effect. This includes procainamide and epigallocatechin-3-gallate, a compound of green tea [12–14].

HIGH-DOSE AZANUCLEOSIDES IN THE TREATMENT OF LEUKEMIA: CLINICAL RESULTS OF AGGRESSIVE COMBINATION CHEMOTHERAPY

5-AZACYTIDINE

Clinical development of 5-azacytidine for leukemia treatment was performed predominantly in the United States. As early as 1976, von Hoff et al. [22] communicated the experience of eight phase II studies of 5-azacytidine-containing regimens, mostly

TABLE 14.1
Drugs with Demethylating Activity Developed for Clinical Use

Drug	Mode of Action	Clinical Trial
5-azacytidine (5-aza-CR)	1) Inhibition of DNA methyltransferase [15]	++
	2) Incorporation into RNA>DNA resulting in disassembly of polyribosomes, defective methylation and acceptor function of tRNA, inhibition of protein production [16]	
5-aza-2'-deoxycytidine (5-aza-CdR, decitabine)	Inhibition of DNA methyltransferase [15]	++
	Incorporation into DNA>RNA and blocking of DNA synthesis resulting in direct toxicity [17]	
pseudoisocytidine (psi ICR)	Inhibition of DNA methyltransferase [15,18]	+
zebularine(1-(beta-D-ribofuranosyl)-1,2-dihydropyrimidin-2-one)	Inhibition of DNA methylation, oral formulation [19]	+
arabinosyl-5-azacytidine (fazarabine)	Inhibition of DNA methylation [20]	+
5-6-dihydro-5-azacytidine (DHAC, dH-aza-CR)	Inhibition of DNA methylation, stable in aqueous solution [15,21]	++
5-fluoro-2'-deoxycytidine (FCdR)	Inhibition of DNA methyltransferase, Dnmt [15]	

Note: ++ = various clinical trials; + = phase I trials.

in relapsed/refractory acute myeloid leukemia, with a substantial proportion of the patients being heavily pretreated. In that meta-analysis, a rate of objective responses of 36% was reported, including 20% complete remissions (CR). In 1987, Glover et al. [23] summarized the experience of 335 patients with 5-azacytidine in leukemia. These included numerous drug combinations, and resulted in an overall complete response rate of 16.7%. In the largest trial reported in that meta-analysis [24], the CR rate in 101 patients with relapsed/refractory AML was 8%.

5-AZA-2'-DEOXYCYTIDINE (DECITABINE)

Development of decitabine in AML and CML blast crisis also demonstrated activity of this drug used either alone or in combination with an anthracycline in relapsed/refractory disease (Table 14.2). The overall activity was comparable to that of high-dose cytarabine. While the total doses of 5-azacytidine were between 400 and 2000 mg/m², with most studies using total doses of 750 mg/m², decitabine was used at somewhat higher median doses. Toxicities with decitabine were also marked at the high doses, including hepatotoxicity, neurotoxicity with reports of coma, and severe mucositis. A trial conducted by the Brazilian group [25] used decitabine at a dose of 90 mg/m² daily over 5 days (total dose 450 mg/m²) combined with daunorubicin as first-line treatment of *de novo* AML. They reported a CR rate of 100%, with a median survival of 15 months, and two patients surviving longer than 3 years (G. Schwartsmann, M. S. Fernandez, personal communication).

TABLE 14.2
Clinical Trials of High-Dose 5-Azacytidine and Decitabine in Hematological Malignancies

Indication	Drug(s)	Patients, n	Total Dose (mg/m²)	Schedule (mg/m²)	Percentage Responses (%, CR, PR)	Median Response Duration/Survival (Months, Range)	Reference
AML, salvage	azaC	14	750–1000	150–200, 5d, i.v.	43 (36/7)		[26]
AML, salvage	azaC	18	750–2000	150–400, 5d, i.v.	39 (17/22)		[27]
AML, salvage	azaC	200	750–2000	150–500, 5d, i.v.	36 (20/16)		[22] (review summarizing 8 studies)
AML, salvage	azaC	50	1500 1500	100, q8h × 15, i.v. 75, q6h × 20, i.v.	30 (20/10)		[28,29]
AML, salvage	azaC	101	750 1000–1500	250, q4h × 3, i.v. 150–300, 5–10d, c.i.	8 (8/0)		[24,29]
AML, salvage	azaC	15	1000	200, 5d	47 (33/14)	28(6–138)/ 35(17–148)	[30]
AML, CML bc, salvage	azaC mitoxantrone	53	600	200, 3d	15 (15/0)		[31]
ANLL (AML), induction	azaC VP16	14	750	50, q8h × 15	(0/0)	4–5/NA	[29,32]
AML, induction	azaC thioguanosine	81	750	150, 5d, c.i.	27 (20/7)		[29,33]
AML, induction	azaC pyrazofurin	27	750	150, 5d, c.i.	26 (11/14)		[29,34]

TABLE 14.2 (CONTINUED)
Clinical Trials of High-Dose 5-Azacytidine and Decitabine in Hematological Malignancies

Indication	Drug(s)	Patients, n	Total Dose (mg/m²)	Schedule (mg/m²)	Percentage Responses (%, CR, PR)	Median Response Duration/Survival (Months, Range)	Reference
AL, induction	azaC daunorubicin cytarabine prednisone vincristine	163	400	50, q12h × 8	72 (72/0)		[29,35]
AML, post-remission induction	azaC azaC + thioguanosine	335	750 1500	150, 5d, c.i. 300 5d, i.v.			[36]
AML, induction	azaC amsacrine	53	800	200, 4d, c.i.	21 (13/9)	7(3.5–13.5)/NA	[29,37]
AML, salvage	azaC	45	250–1000	50–200, 5d every 2wk, c.i.	41 (24/9)	3(1–20)/NA	[26]
CML, bc	azaC VP16	27	750	150, 5d, i.v.	58 (3/55)	NA/5(1–10.5+)	[27]
CML, bc	azaC mitoxantrone	40	750	150, 5d, i.v.	23 (13/5)		[38]
AML, ALL salvage	DAC	22	NA	0.75–80* (mg/kg), 12–44h, c.i.	14 (9/5)	0.5(0–1.5)/NA	[39]
AML, salvage	DAC	27		37–81*, 36–60h, c.i.	85 (22/15)	1/5**(1–15)	[6]
AML, MDS, CML	DAC	19	135–910	15–90, q8h 3d, i.v.	70 (15/30)	12(6–58+)/ 19(7–64+)	[40]
AML, salvage	DAC amsacrine	10	750–1500	125–250, 6d, i.v.			[41]
AML	DAC	12	270–360	90–120, 3d, i.v.	33 (25/8)		[42]

TABLE 14.2 (CONTINUED)
Clinical Trials of High-Dose 5-Azacytidine and Decitabine in Hematological Malignancies

Indication	Drug(s)	Patients, n	Total Dose (mg/m²)	Schedule (mg/m²)	Percentage Responses (%, CR, PR)	Median Response Duration/Survival (Months, Range)	Reference
AML, salvage	DAC + m-amsacrine or idarubicin	22	1500	125, q12h 6d, i.v.	68 (59/9)	4/NA	[43]
AML, untreated, induction	DAC daunorubicin	8	450	90, 5d, i.v.	100 (100/0)		[25]
CML, bc and accelerated phase	DAC	37	750–1000	75–100, q12h 5d, i.v.	37 (5/5)		[44]
AML, salvage	DAC + m-amsacrine or idarubicin	63	1500	125, q12h 6d, i.v.	(36.5/NA)		[45]
CML, bc	DAC	31	500–1000	50–100, q12h 5d, i.v.	25 (6/0)	NA/29	[46]
AL, CML***	DAC	14	1000–1500	100–150, q12h 5d, i.v.	57	6(0–40.5+)/NA	[47]
AML, CMML, ALL, CML, conditioning	DAC busulfan cyclophosphamide	23	400–800			17.2	[48]
CML	DAC	130	500–1000	50–100, q12h 5d, i.v.	42		[49]

Note: i.v. = intravenously; c.i. = continuous infusion; bc = blast crisis; AL: acute leukemia; NA: not given; * = mg/kg; ** = mean; *** = followed by second transplant.

LOW-DOSE AZANUCLEOSIDES AS NOVEL TREATMENT OF PRELEUKEMIA (MDS) AND ACUTE MYELOID LEUKEMIA (AML)

In contrast to the development of both azanucleosides in a cytarabine-like application, i.e., at high doses reaching the maximally tolerated range or as part of an antileukemic polychemotherapy, the low-dose approach with both drugs followed a quite different path. Pinto et al. (1989) in their pioneering studies used total doses of decitabine that were significantly lower [40]. In their first series of 10 patients with high-risk myelodysplastic syndrome ([50], Table 14.3) they reported a 50% overall response rate with 40% complete remissions, using a 3-day infusional schedule (continuous infusion, or 4-hour infusions followed by 4-hour rest periods).

5-AZACYTIDINE

A 7-day schedule was developed in the United States for 5-azacytidine by Silverman and coworkers, also resulting in substantial number of complete remissions when given intravenously [51]. The Cancer and Leukemia Group B (CALGB) conducted a phase II study using the same total dose of 525 mg/m² but given subcutaneously (also over 7 days repeated every 4 weeks). This study ([51], Table 14.3) yielded a complete response rate of 12%, with overall responses of 53% (including 27% responses not fulfilling the criteria for partial remission; i.e., single lineage improvement such as correction of thrombopenia or anemia). Based on the good tolerance of this schedule, the same group initiated a phase III comparative trial.

In this trial (CALGB 9221), recently reported by Silverman et al. [52], 191 patients were randomized to either subcutaneous 5-azacytidine for 7 days, repeated every 28 days ($n = 99$), or supportive care ($n = 92$). A crossover to 5-azacytidine treatment was allowed after 4 months in case of disease progression (n = 46). The majority of patients suffered from RAEB or RAEB-t, but the IPSS risk score [53], available for 81/191 patients, was mostly intermediate-1 (45%) or intermediate-2 (27%). Significant differences were noted in the rate of complete remissions (7% vs. 0%), partial remissions (16% vs. 0%) and "improvement" (37% vs. 5%) between the 5-azacytidine-treated group and the group receiving only supportive care. Median duration of these responses was 15 months. Highly significant differences were seen in the patients' times on-study before exiting the trial because of failure. By intention-to-treat, transformation to AML was 2.8-fold more frequent in the supportive-care group than in the 5-azacytidine group ($p = 0.003$), suggesting that 5-azacytidine may prolong time to transformation to acute leukemia.

In this trial, the median overall survival was 20 months in patients randomized to immediate 5-azacytidine treatment, compared to 14 months in those assigned to supportive care (difference not statistically significant). To compensate for the confounding factor of more than half of these patients crossing over to 5-azacytidine, a 6 month "landmark analysis" was performed, resulting in a statistically significant difference in survival from the landmark between patients on the treatment arm (18 months), and those receiving 5-azacytidine after the landmark or never (11 months). A quality-of-life analysis showed that patients treated initially with 5-azacytidine

TABLE 14.3
Low-Dose Azanucleosides (5-Azacytidine, Decitabine) in Myelodysplastic Syndromes and Acute Myeloid Leukemias

Indication	Drug(s)	Patients, n	Median Age (Years, Range)	Total Dose (mg/m²)	Schedule (mg/m²)	Percentage Responses (%, CR, PR, HI)	Median Response Duration/Survival (Months, Range)	Reference
AML, salvage	azaC	11	55 (36–78)	525	75, 7d, c.i.	0 (0/0/0)	2/NA	[59]
MDS	azaC	15	64 (31–76)	140–490	10–35, 14d, c.i.	23 (0/0/23)		[60]
MDS	azaC	43	65	525–1050	75–210, 7d, s.c.	49 (12/25/12)	14.7 (CR+PR)/13.3	[51]
MDS	azaC	68	66 (23–82)	525	75, 7d, s.c.	53 (12/15/27)	17.3/NA	[61]
MDS	azaC	92**	NA (27–91)	525	75, 7d, s.c.	61 (13/29/14)	NA(2–30+)/NA	[62]
AML	azaC + phenylbutyrate	18	67 (48–81)	250–525	5–10d, s.c.	22 (0/11/11)	NA	[63]
MDS	azaC/ BSC*	98** 92**	69 (31–92)	525	75, 7d, s.c.	60 (7/16/37) 5 (0/0/5)	15/20 2/14	[64]
MDS	azaC	46	68	525	75, 7d, s.c.	39 (0/0/39)	7/NA	[65]
AML	DAC	12	NA	270–360	90–120, 3d, i.v.	33 (25/8/NA)		[42]
MDS	DAC	10	68 (60–78)	135–150	45, 4h × 3d, i.v.; 50, 3d, c.i.	50 (40/10/0)	11/NA	[66]
MDS	DAC	29	72 (58–82)	120–225	40–75, 3d, c.i.	54 (29/18/7)	7.3/10.5	[55]
MDS	DAC	66	68 (38–84)	135	45, 4h × 3d, i.v.	49 (20/4/25)	7.3/15	[67]
MDS	DAC	169	70 (38–89)	135	45, 4h × 3d, i.v. ***	49 (23/12/14)	9/15	[56]
AML/MDS, salvage; CML, ALL	DAC	50	60 (2–84)	50–200	5–20, 10–20d	28 (18/2/8)	8(4–59+)/44	[58]
AML	DAC	4	77 (62–79)	135	45, 4h × 3d i.v.	50 (2/0/0)	6/9(5–27)	[68]

Note: i.v. = intravenously; c.i. = continuous infusion; s.c. = subcutaneously; * = possibility of cross-over if progressive disease; ** = including patients with >30% blasts (secondary AML according to FAB); *** = also by c.i.

had a significant improvement over time in fatigue, dyspnea, physical functioning, and psychological distress, compared to those receiving supportive care only [54].

5-Aza-2′-Deoxycytidine (Decitabine)

Wijermans and coworkers performed several studies, using the 3-day schedule of decitabine pioneered by Pinto and colleagues, to treat mostly elderly patients with myelodysplasia. The first study of 29 patients (median age 72 years) resulted in an overall response rate of 54%, including 29% complete remissions [55]. A cumulative analysis of 169 patients treated in a total of three phase II studies confirmed this outcome, with an overall response rate of 49%, including 23% complete responses [56]. Interestingly, among patients presenting with an informative cytogenetic abnormality, major cytogenetic responses during continued low-dose decitabine treatment were observed in 31% [57] and the cytogenetic response rate as well as survival was better in the subgroup with "high-risk" cytogenetic abnormalities according to the IPSS score [53] compared to patients with "intermediate risk" cytogenetic abnormalities.

At present, the unusually favorable response of patients with complex karyotype or abnormalities of chromosome 7 to this drug is unclear. However, it was the basis for shaping inclusion to a randomized phase III trial tailored upon the risk strata of the IPSS of 11 to 30% bone marrow blasts and/or high-risk cytogenetics. This trial, activated in Europe, compares survival in patients randomized to continuous low-dose decitabine (maximum treatment duration 12 months) to patients randomized to supportive care only. The trial, by virtue of its design not allowing a crossover can study overall survival as a primary endpoint. A similar trial, also using the 72-hour infusional schedule with 135 mg/m^3 total dose, has finished recruitment in the United States. Whereas the European trial includes patients with the intermediate-2 and high-risk profile, the U.S. trial also includes patients with MDS of intermediate-1 risk.

Recently, Issa and coworkers [58] have used doses of decitabine ranging from 50 to 300 mg/m^2 given over 1 hour intravenously over a time span of 10 to 20 days in (mostly pretreated) patients with relapsed/refractory AML or MDS, also including five patients with CML, and one patient with ALL. Interestingly, they observed the highest response rate (five/six patients) in the group treated with a total dose of 150 mg/m^2, whereas at high doses, the efficacy was lower. This scheduling also differs from the intravenous 3-day scheduling used in both decitabine phase III studies as it includes the possibility of giving decitabine on an outpatient basis.

COMBINATION OF LOW-DOSE DNA DEMETHYLATING AGENTS WITH INHIBITORS OF HISTONE DEACETYLASES: CLINICAL RESULTS

5-Azacytidine has also been employed in combination with sodium phenylbutyrate (PB), an aromatic fatty acid with cytostatic and differentiating activity against malignant myeloid cells with an ID$_{50}$ of 1 to 2 mM. Several mechanisms have been proposed for its clinical activity. At low concentrations (0.25 to 0.5 mM), it has an effect on histone acetylation via an inhibitory activity on histone deacetylase, thus inducing histone H3 and H4 acetylation [69]. PB, like other HDAC inhibitors, synergizes with demethylating agents [70–72].

At the Johns Hopkins Cancer Center, patients with MDS ($n = 11$) and AML ($n = 16$) were treated with PB as a 7-day continuous infusion, repeated every 28 days, in a phase I dose-escalation study [73]. The maximum tolerated dose was 375 mg/kg/day, with dose-limiting, reversible neurocortical toxicity. The median steady-state plasma concentration at this dose was 0.29 to 0.16 mM. Although no patients achieved complete or partial remission, four patients achieved hematological improvement. Monitoring of the percentage of clonal cells over the course of PB administration showed that hematopoiesis remained clonal [74]. Hematological response was often associated with increases in colony growth *in vitro*. Thus, at the plasma concentrations achieved, PB influences the expression of at least one surrogate gene involved in hematopoiesis.

The outstanding toxicity profile of PB and the synergistic effect of histone deacetylation and demethylating agents in reactivating silenced genes encouraged clinical studies on the combination of PB and demethylating agents in hematological diseases. A treatment scheme entailing subcutaneous injections of 5-azacytidine for 7 consecutive days (75 mg/m²/day), similar to the CALGB schedule, followed by 5 days of intravenous PB (200 mg/kg/day), was reported [75]. Six myelodysplasia/secondary AML patients received at least one cycle of therapy (range, 1 to 3). Reduction in bone marrow blast counts, as well as increased myeloid maturation was observed in four patients; one patient with leukemia who relapsed following bone marrow transplantation (BMT), had a complete elimination of bone marrow blasts after one cycle of therapy. An increase in histone acetylation was consistently detected in peripheral blood and bone marrow samples collected after PB administration. Treatment was relatively well tolerated, with mild adverse reactions including fatigue, nausea, vomiting, and local tenderness at injection sites, as well as somnolence and drowsiness associated with PB.

A second study with sequential administration of 5-azacytidine and PB to re-express transcriptionally silenced genes was also initated at Johns Hopkins in patients with MDS and AML. The initial 5-azacytidine dose was also 75 mg/m²/day subcutaneously, was given for 5 days, followed by PB at 375 mg/kg/day by intravenous, continuous infusion, days 5 to 12. Dose de-escalation to determine the minimal 5-azacytidine dose associated with significant demethylation was performed. Eleven patients have been treated in a total of 39 courses. The combination was well tolerated, and two patients had significant hematopoietic improvement. Increases in histone acetylation by this therapy were detected within 4 hours of initiation of PB infusion, and persisted throughout the infusion. In nine patients, sequential measurements of *p15^{INK4B}* promoter methylation by a newly developed PCR-based assay were performed. All had measurable hypermethylation of the *p15* promoter, which was higher in patients with AML or RAEB-t compared to patients with low-risk MDS. In three patients during 5-AC/PB treatment, *p15* methylation levels decreased. Baseline methylation density did not predict for the extent of demethylation in response to 5AC/PB (reviewed in [74]). Both studies successfully demonstrate that the sequential administration of a first generation demethylating agent and HDAC inhibitors is feasible, and gives preliminary evidence of an effect on the methylated targeted gene promoter.

Among the novel HDAC inhibitors being developed for clinical use, several have been studied in refractory/relapsed hematologic malignancies. This includes depsipeptide, SAHA, LAQ814, and MS-275. Results of a phase I study of MS275 in 25 patients has recently been reported [76]. While no complete remissions were noted, this oral compound induced bone marrow partial remissions and hematologic improvement in four patients (18%). Toxicities included fatigue, anorexia, and electrolyte disturbances. Valproic acid, which has recently been ascribed HDAC activity, has been administered, alone or in combination with all-trans retinoic acid, to mostly elderly patients with untreated MDS or AML [77]. While the treatment was well tolerated and interesting clinical effects were seen, no prolonged disease control was achieved, thus prompting future studies with other combinations. With both oral HDAC inhibitors, active and rapid acetylation of histones H3 and H4 was observed in normal peripheral blood lymphocytes of the patients.

CLINICAL RESULTS OF DNA DEMETHYLATING AGENTS IN SOLID TUMORS: PHASE I/II TRIAL RESULTS

Promising *in vitro* activity of 5-azacytidine and decitabine on cell lines and in rodent tumor models has prompted a number of clinical phase I/II trials of both compounds in solid tumors. Most of these were performed in the 1980s and yielded only a low percentage of responses. The clinical results are summarized in Table 14.4. Side effects were similar to those observed in patients with hematologic malignancies.

Momparler et al. recently performed a phase I/II trial with decitabine (given intravenously over 8 hours) at doses ranging from 200 to 660 mg/m^2 to 15 patients with metastatic non-small-cell lung cancer. The major toxicity noted was hematopoietic in these chemotherapy-naive patients, necessitating recovery periods for up to 6 weeks before repetition of treatment. Four of nine assessable patients had stable disease for at least 6 months, with one of them living for more than 5 years [78]. The median survival of all patients was 6.7 months.

Schwartsmann and coworkers performed a phase II trial of cisplatinum (30 to 40 mg/m^2 on 1 day) combined with decitabine (50 mg/m^2 for 3 consecutive days, repeated every 21 days) in 25 chemotherapy-naive patients with advanced squamous cell carcinoma of the cervix [79]. A median of three courses per patient was administered (range 1 to 8), with grade III/IV neutropenia being the dominant toxicity (79% of patients had received prior radiotherapy). Eight of 21 evaluable patients (38%) achieved a partial response and stable disease was attained in 5 of 21 cases (24%). Interestingly, objective responses were more frequent in metastatic lesions at nonirradiated sites.

TABLE 14.4
Treatment of Solid Tumors with 5-Azacytidine and Decitabine

Indication	Drug(s)	Patients, n	Total Dose (mg/m²)	Schedule (mg/m²)	Response (%)	Reference
Head and neck	DAC	27	225	75, q8h 1d, i.v.	0	[80]
NSCLC	DAC	9	200–660	200–660, 1d, i.v.	44	[78]
NSCLC	DAC/cisplatin	14	180	3d, i.v.	21	[81]
Malignant mesothelioma	DHAC	29	7500	1500, 5d, c.i.	17	[82]
Malignant mesothelioma	DHAC	41	7500	1500, 5d, c.i.	17	[83]
Gastrointestinal cancer	azaC	27	ND	ND	4	[84]
Colorectal carcinoma	DAC	42	225	75, q8h 1d, i.v.	0	[80]
Breast cancer	azaC	11	300–700	37.5–87.5, 8d, i.v.	63	[85]
Breast cancer	azaC	31	600	10d, i.v.	6.4	[85]
Breast cancer	azaC	4	275–850	27.5–85, 10d, s.c.	25	[86]
Ovarian carcinoma	DAC	24	225	75, 3d, i.v.	8	[87]
Cervical carcinoma	DAC	15	225	75, q8h 1d, i.v.	0	[88]
Cervical carcinoma	DAC/cisplatin	21	150	50, 3d, i.v.	62	[79]
Prostate cancer	DAC	14	225	75, q8h 1d, i.v.	17	[89]
Malignant melanoma	azaC	30	1000	100, 10d, s.c.	0	[90]
Malignant melanoma	DAC	18	225	75, q8h 1d, i.v.	7	[80]
Malignant melanoma	DHAC	40	5000	5000, 1d, c.i.	20	[91]
Different solid tumors	azaC	177	16 mg/kg	1.6 mg/kg, 10d, i.v., c.i.	17 (breast cancer)	[92]
Different solid tumors	DAC	21	50–300	25–100, q12h/q8h 1d, i.v.	5	[93]
Different solid tumors	DAC/cisplatin	21	135–360	45–120, 3d, i.v.	14	[81]
Different metastatic solid tumors	DAC	19	60–120	20–40, 3d, c.i.	0	[94]

Note: NSCLC = non-small-cell lung cancer; i.v. = intravenously; c.i. = continuous infusion; s.c. = subcutaneously; ND = not done.

LOW-DOSE AZANUCLEOSIDES AS NOVEL
TREATMENT IN HEMOGLOBINOPATHIES

Van der Ploeg and Flavell [95] first demonstrated hypomethylated gene regions upstream of the duplicated gamma-globin μ-locus (encompassing the Gμ and Aμ genes) in fetal liver cells. Remethylation of these loci has occurred in adult human bone marrow cells, in which the μ-globin genes are not expressed. This seminal discovery provided a reason to explore the pharmacologic stimulation of fetal hemoglobin (HbF) production, through induction of gamma-globin messenger RNA (mRNA) expression with demethylating agents, in patients with severe beta-thalassemia. In anemic baboons, short-term 5-azacytidine administration resulted in a marked transient enhancement of HbF synthesis [96].

In 1982, Ley et al. reported clinical results on a patient with severe beta-thalassemia [97]. 5-azacytidine was administered by continuous infusion at 2 mg/kg/day for 7 days (Table 14.5). Following infusion, a rise in hemoglobin (Hb) levels from 8 g/dl to 10.8 g/dl and a more than fourfold increase in reticulocytes were attained. Hb values remained above pretreatment levels for nearly 5 weeks without transfusion. Side effects included moderate nausea, vomiting, and mild leukopenia (noted 3 weeks after end of treatment). *In vitro* studies revealed a sevenfold induction of μ-globin mRNA, with a marked increase of both Gμ- and Aμ-chain synthesis. Transient hypomethylation of the gamma-globin gene locus coincided with augmented HbF production.

Charache et al. treated a 32-year-old patient with sickle cell anemia and severe transfusion requirements with 5-azacytidine infusions [98]. Also in this patient, a rise in HbF, reticulocytes and Hb levels was obtained and, upon decrease to baseline values, could be reinduced. Since these pioneering studies, further reports have described similar effects in hemoglobinopathy patients (Table 14.5). In 1993, Lowrey and Nienhuis reviewed the long-term responses and outcome of three thalassemia patients. All three became transfusion-independent (two of them for more than a year) with repeated courses of intravenous 5-azacytidine. One patient had a remarkable improvement in cardiac function, with effective iron depletion only after the addition of 5-azacytidine. None of the three patients developed a malignancy [99].

The group of DeSimone explored the effects of decitabine in patients with sickle cell anemia unresponsive to hydroxyurea (HU) [100]. In a preliminary report, they noted a threefold increase in HbF in all nine patients treated with very low doses of intravenous decitabine (starting dose: 0.15 mg/kg/day) for 5 days, repeated once during the course of 2 weeks. Maximum levels of HbF were reached within 4 weeks and were sustained for at least 2 more weeks. The only side effect noted was mild, reversible neutropenia. They concluded that, with the efficacy of this low-dose, short-term schedule in a patient group not responding to HU, further increases in HbF might be attained upon dose escalation. Very recently, this group reported remarkable improvement of hemolysis and hemoglobin levels in eight decitabine treated patients with sickle cell disease who had received prior hydroxyurea [101]. Decitabine was given subcutaneously 1 to 3 times per week in two cycles of 6 weeks duration (0.2 mg/kg/day). The biological correlate of the hematologic improvement was induction of hemoglobin F in all cases. Further biological effects of this treatment, which was

TABLE 14.5
Clinical Effects of 5-Azacytidine and Decitabine in Severe β-Thalassemia and Sickle Cell Anemia

Drug	Hemoglobinopathy	Patients, n	Total Dose (mg/kg)	Schedule (mg/kg)	Maximal HbF (%)	Maximal Increase in Hb (g/dl, range)	Demethylation	Reference
azaC	β+ thalassemia	1	14	2, 7d, c.i.	20.8	2.8	Gμ, Aμ, ε	[97]
azaC	Sickle cell anemia	1	90	30, 3d, c.i. & s.c.	8.9	3.0	Global, Gμ, Aμ, Y rpt	[98]
azaC	β+ thalassemia (2), sickle cell anemia	4	14	2, 7d, c.i.	21.4	1.9 (1.2–2.9)	Gμ, Aμ,	[102]
azaC	Sickle cell anemia	4	30–70	2, 15–35d, s.c.	13.6	3.0 (1.2–3.5)	ND	[103]
azaC	Sickle cell anemia	7	14 7.5	2, 7d, c.i. 1.5, 5d, c.i.	ND.	1.4 (0.2–2.5)	Gμ, Aμ,	[104]
azaC	β° thalassemia	1	10	2, 5d, c.i.	ND	3.0	ND	[105]
azaC	β+ thalassemia (2) β° thalassemia	3	4–8	1–2, 4d, c.i.	76	3.1 (2.9–4.4)	ND	[99]
DAC	Sickle cell anemia	9	1.5	0.15, 10d, i.v.	9.1	NA	ND	[100]
DAC	Sickle cell anemia	8	1.5–3.0	0.15–0.3, 10d, i.v.	13.5	1.0	ND	[106]
DAC	Sickle cell anemia (6), α-thalassemia	7	3	0.3, 10d, i.v.	18.4	2.5	ND	[107]
DAC	Sickle cell anemia	8	1.6–4.8	8–24d, s.c.	20.4	2 (7.6–9.6)	G	[101]

Note: i.v. = intravenously; c.i. = continuous infusion; s.c. = subcutaneously; ND = not done; G = gamma globin; A = alpha globin.

very well tolerated, included gamma-globin promoter demethylation, peripheral blood platelet increases, and an increase of megakaryocytes and erythroid precursor cells in the bone marrow in the absence of hypoplasia. These results warrant prolonged drug exposure, since sustained HbF and total Hb increases may be possible to achieve.

CLINICAL EFFICACY AND TRANSLATIONAL STUDIES OF DEMETHYLATING AGENTS: BIOLOGICAL EFFECT *IN VIVO*

TARGETING OF HYPERMETHYLATION OF CELL CYCLE REGULATORS IN HUMAN MALIGNANCY BY DNA DEMETHYLATING AGENTS

Progression through the cell division cycle is regulated by an intricate interplay of cyclin-dependent kinases (CDKs) and cyclins (Figure 14.1). CDK inhibitors (CDKIs) regulate the activity of CDK/cyclin complexes and are able to suppress proliferation. CDK inhibitors are classified into two families: the INK4-family and the CIP/KIP-family. The CIP/KIP family members, $p21^{WAF1/CIP1}$, $p27^{KIP1}$, and $p57^{KIP2}$, can bind to and block the activity of all CDK/cyclin complexes [108–112], whereas the INK4-inhibitors $p15^{INK4B}$, $p16^{INK4A}$, $p18^{INK4C}$, and $p19^{INK4D}$ inhibit only the activities of cyclin/CDK4 and cyclin/CKD6 [113–117]. Expression of many cell cycle regulatory genes is frequently suppressed in human malignancies by methylation of their promoter regions [reviewed by 16,118–120]. This promoter hypermethylation results in silencing of gene expression [121]. Promoter silencing of all CKIs except $p19^{INK4D}$ and $p21^{WAF1/CIP1}$ has been detected in a plethora of human cancers and leukemias [122–128]. $p15^{INK4B}$ is frequently hypermethylated in leukemias and myelodysplastic syndromes (MDS), whereas $p16^{INK4A}$ hypermethylation is frequently found in solid tumors, multiple myeloma, and lymphomas [129,130]. $p18^{INK4C}$ is hypermethylated in a subset of Hodgkin's lymphoma [122]. Methylation of these genes may have prognostic relevance, which needs to be assessed in prospective clinical studies.

BIOLOGICAL EFFECTS OF DECITABINE ON PROMOTER METHYLATION AND PROTEIN EXPRESSION

The demethylating agents 5-azacytidine and 5-aza-2′-deoxycytidine (decitabine) can reverse epigenetic silencing by inhibiting DNA-methyltransferases, and reconstitute gene expression through hypomethylation of CpG islands within promoter regions [131,132]. Decitabine was first reported as a differentiation-inducing drug [133], but in addition it can exert cytotoxic effects [17] and may induce apoptosis [134,135]. Data obtained from *in vitro* experiments raised high hopes for a less toxic treatment option of leukemias (*vide supra*). Indeed, several clinical trials with decitabine for treatment of several myeloid neoplasias at doses with minimal nonhematologic toxicity (plasma levels 0.1 to 0.5 µM), particularly in elderly patients, show promising results (Table 14.3) [120,131].

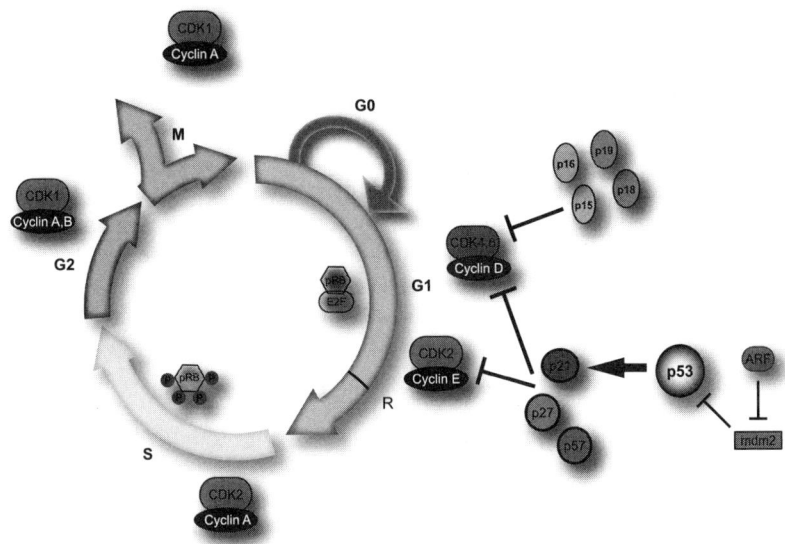

FIGURE 14.1 (**See color insert following page 114**) Schematic diagram of the cell division cycle in human cells. During G1 phase, cells increase cell size and synthesize new proteins necessary for further progression through the cell cycle. Cells can either progress through G1 and enter S phase, or withdraw from the cell cycle (G0). During S phase, the complete genome is duplicated once. In G2, DNA replication is completed and cells prepare to enter mitosis (M phase). During M phase, a bipolar mitotic spindle is formed, ensuring equal chromosome segregation. CDK2, CDK4, and CDK6 in combination with cyclin D and E, drive the cell cycle progression. The CDK inhibitors (INK4 and CIP/KIP) regulate and can also block their cell cycle promoting activities.

Hypermethylation of the p14^{ARF} locus has been detected in various cancer cell lines and primary human cancers, and can be reversed by decitabine treatment [136], but clinical significance of these findings awaits further confirmation. The importance of p15^{INK4B}-repression during leukemogenesis has been well documented in gene targeting studies, in which the complete or partial p15^{Ink4b} knockoutcorrelates with increased susceptibility to transformation [137,138]. Using retrovirus-induced myeloid leukemia in mice, Wolff et al. frequently detected hypermethylation of the 5′ CpG island of the Ink4b gene, that could be reversed following treatment with decitabine. In this experimental setup, targeted deletion of one allele of the p15^{Ink4b} gene increased the susceptibility to retrovirus-induced myeloid leukemia [138]. This work confirms our report on MDS patients, wherein the clinical efficacy of decitabine correlates with hypomethylation of the p15^{INK4B} promoter [139]. Daskalakis et al. studied p15^{INK4B} methylation and expression in bone marrow cells from patients with myelodysplastic syndrome before, during, and after treatment with low-dose decitabine, using the highly sensitive and specific Ms-SNuPE assay, as well as bisulfite sequencing [139]. Sixty-five percent of the patients showed p15^{INK4B} promoter methylation prior to treatment, and methylation was reduced during the initial courses of treatment, correlating with increased p15^{INK4B} expression. A recent report, using a different drug schedule and methylation detection assay, showed that low dose decitabine was effective for treatment of myeloid malignancies but did not correlate with p15^{INK4A} demethylation [58].

Treatment with decitabine can also restore p16^{INK4A} expression in various *in vitro* models of solid and hematological malignancies, but so far does not necessarily correlate with tumor response to treatment [125,140,141] (reviewed by 120]). In an analysis of human lung cancer cell lines, promoters of the genes coding for p16^{INK4A}, but not p19^{INK4D} showed CpG island methylation. Interestingly, treatment of cells with decitabine induced a dose- and duration-dependent increase in p16^{INK4A} and p19^{INK4D} expression. In this study, elevated p19^{INK4D} was not attributed to changes is methylation but to inhibition of histone deacetylation [142].

Effect of Decitabine on CIP/KIP Expression

Induction of p21$^{WAF1/CIP1}$ expression has been frequently observed upon treatment with demethylating agents [141,143]. Data about p21$^{WAF1/CIP1}$ promoter methylation is incongruent, and so is the question whether the observed p21$^{WAF1/CIP1}$-induction is due to hypomethylation after decitabine treatment or a more indirect effect. In one study on myeloma cell lines, the increased protein expression precedes upregulation of p16^{INK4A} and could be observed at lower doses (2 to 4×10^{-8} vs. 10^{-6} to 10^{-7} M) [141]. In a study using gastric carcinoma cell lines, only p57^{KIP1} was shown to be hypermethylated, whereas p21$^{WAF1/CIP1}$ and p27^{KIP1} were not [144]. In a model using rat and mouse pituitary tumor cell lines, p27^{KIP1} was hypermethylated, and treatment with decitabine resulted in promoter hypomethylation, leading to restoration of p27^{KIP1} mRNA and protein expression [145].

Methylation and Acetylation

Chromatin folding is regulated by distinct modifications of histone proteins. This mechanism is tightly controlled by histone deacetylase (HDAC)-mediated histone deacetylation and results in transcriptional repression. Transcriptional activation by histone hyperacetylation and subsequent chromatin remodeling, thus allowing access of the transcriptional machinery to gene promoters, is mediated by histone acetyltransferases (HAT). The balance between HAT and HDAC activity is frequently deregulated in malignancies. Treatment of leukemia cell lines with HDAC inhibitors results in gene reactivation and subsequent differentiation or apoptosis [146]. A number of compounds with HDAC inhibitory activity have come to clinical attention, and several of these are being developed for the treatment of hematopoietic neoplasias [147].

Decitabine does not only inhibit DNA methyltransferases, but also can induce histone acetylation in bladder cancer cells [148]. Treatment of AML patients with 5-azacytidine alone resulted in acetylation of genes in five of six patients. This effect could be recapitulated in leukemia cell lines upon treatment with decitabine, an effect that is increased by addition of specific HDAC inhibitors [143]. Combination therapy of HDAC inhibitors with decitabine increased the proapoptotic effects of HDAC inhibitors in human lung cancer cells [149]. As therapeutic targets, both gene hypermethylation and hypoacetylation of histones are models for which therapeutic interventions are being developed [70]. Combination treatment of demethylating agents and HDAC inhibitors may help reversing epigenetic gene silencing, thus sensitizing cells for growth and differentiation inducing signals.

DIFFERENTIATION VS. APOPTOSIS

Blast clearance in patients is frequently observed upon treatment with demethylating agents (reviewed by [120]). Whether this is due to a direct, cytotoxic effect, a proapoptotic effect, or differentiation of the transformed clone is still a matter of debate. There are conflicting data about the cytotoxicity of decitabine and the involvement of the tumor suppressor protein p53. DNA damage and activation of subsequent cellular pathways has been observed after incorporation of deoxy-azanucleoside into the DNA strand [17]. Induction of $p21^{WAF1/CIP1}$ and cell cycle arrest, but no increase in apoptosis, were reported upon treatment of human myeloma and lymphoblastic cell lines with decitabine [141]. In colon cancer cell lines, decitabine treatment resulted in DNA-damage-induced p53 activation with subsequent $p21^{WAF1/CIP1}$ activation and induction of apoptosis [135]. Contrasting data have been reported recently in a model system using primary mouse embryonic fibroblasts (MEF) [134]. In this system, only MEFs deficient for p53 undergo apoptosis after treatment with decitabine, whereas p53 wild type MEFs did not. One possible mechanism for p53-dependent induction of apoptosis without direct DNA damage could be that $p14^{ARF}$ hypomethylation leads to increased $p14^{ARF}$ expression and to derepression of p53 through increased degradation of mdm2 [150]. Thus, functional p53 can accumulate, and cells undergo apoptosis.

Work from our laboratory and others suggests that leukemic blasts treated with decitabine do not solely undergo apoptosis, but also differentiate. Upregulation of MHC I on leukemic blasts from patients with AML treated with decitabine was the first report linking decitabine to induction of differentiation [133]. Upregulation of CD11c, CD14, CD16, and CD34 downregulation may reflect differentiation or maturation in myeloid blasts [151]. Yet another hypothesis is that decitabine can exert immune modulatory effects. Upregulation of immunogenic molecules and cancer antigens with improved immune defense has been observed in a recent study [152]. This study examined the expression of cancer testis antigens (CTAs) on myeloid blasts during treatment with decitabine. The authors noted a lack of expression of three CTAs (MAGE1, SSX, NY-ESO-1) in 9 of 11 patients. Following decitabine treatment, these genes were expressed in 8 of 9 cases. In all studies, peripheral blood leukemic blasts were amenable to analysis, making gene expression changes caused by emergence of normal hematopoietic cells unlikely.

Taken together, the promising results from *in vitro* studies and the molecular biology behind them translate nicely into results obtained from various clinical trials with 5-azacytidine and decitabine. Future studies and clinical trials using genomic screens and high throughput analyses of tumors will help to address the question of what the contribution of hypermethylation and histone acetylation to carcinogenesis and tumor progression is, and to further identify molecular targets of decitabine [153,154].

SUMMARY AND OUTLOOK

The clinical development of azanucleosides in cancer treatment can be divided into two periods and principal approaches. One is the use of decitabine and 5-azacytidine in a cytarabine-like fashion, with doses requiring often prolonged hospitalization due to significant hematologic and nonhematologic toxicity. The other uses either drug at low doses in order to achieve biological effects, such as DNA demethylation, rather than maximally tolerated cytotoxicity. This approach is also designed to allow the patient a maximum of remaining lifetime outside the hospital by virtue of a favorable toxicity profile. However, since complete responses to these low-dose schedules in myeloid neoplasias and hemoglobinopathies do not occur in every patient thus treated, continued investigation of the possible target genes of their DNA demethylating activity appears warranted in order to further improve drug scheduling. In addition, identification and validation of ubiquitous marker genes, such as imprinted genes and globin genes which may serve as molecular markers of a minimal demethylating activity of these drugs, would be an important step. Presently, investigation of an additional activity of azanucleosides in inhibiting also histone deacetylases, and modifying histone methylase action is ongoing.

Better knowledge of potential target genes of demethylating agents (and drugs acting on histone modification) will be very useful also in designing rational drug combinations, e.g., with differentiation-inducing signals (retinoids, vitamin D_3 analogues, etc.), growth signals (colony-stimulating factors), and growth inhibitory signals (TGFβ, interferons). Furthermore, an important line of research investigating sensitization of tumors to cytostatic drugs such as cisplatinum and derivatives may result in rational drug combinations of demethylating agents with classic cytotoxic chemotherapy regimens [155].

The broad and growing knowledge of "methylator" phenotypes in different tumors, and of the effects of low doses of azanucleosides in cell line models of epithelial and mesenchymal malignancies, promises to yield a rationale for expansion of clinical trials of azanucleosides in a variety of solid tumors. Of particular interest are malignancies in which a T-cell-mediated immune response may confer clinical benefit. The results demonstrating upregulation of cancer testis antigens and other molecules recognized by T cells point to a role of demethylating agents also in modification of an immune response. Cancers such as renal cell carcinoma, possibly malignant melanoma, and colorectal cancer in addition to myeloid leukemia, may be promising targets. In a corollary to this concept, modification of the efficacy of allogeneic donor lymphocyte infusions after relapse from allogeneic stem cell transplantation may hypothetically be achieved by upregulation of such molecules on residual blasts in these patients [156].

De-escalating drug schedules of azanucleosides to doses that target cancer-associated hypermethylation in malignant cells or silenced hemoglobin genes in patients with hemoglobinopathies are presently being developed. Concerns that reactivation of genes whose expression is undesirable need to take into consideration that this pattern of silencing is not likely in a malignant cell, because inactivation of an oncogene during evolution of the tumor cell does not appear as an epigenetic event that would reasonably yield a growth advantage. CpG island methylation is infrequent in normal cells, making tumor hypermethylation a target for pharmacological therapy, which is both specific and distinct from conventional cytotoxic activity. This is supported by the limited degree of expression changes when normal cells are treated with low doses of demethylating agents. In theory, major targets of reactivation would be X-chromosome-linked genes in females and genes that are imprinted. However, the expression pattern of both classes of genes may be most prevalent during embryogenesis [16]. Of note, in several mouse models targeted demethylation has resulted in a decrease of malignancy or prolongation of tumor development [71,157,158].

In vivo evidence for mutagenic or carcinogenic effects of continued hypomethylation or other mechanisms relieving gene silencing could hypothetically be derived from congenital, lifelong DNA methylation disturbances resulting in a measurable degree of hypomethylation. A paradigm for this situation is provided by the so-called ICF syndrome. In this rare syndrome of immunodeficiency, facial anomalies and centromeric instability, in which marked, global demethylation due to mutation of Dnmt3b [159,160] has been demonstrated, no increase in malignancies has been reported [161]. The long-term exposure to low doses of 5-azacytidine in three patients with hemoglobinopathies, as reported by Lowrey et al. [99] did not include development of neoplasia in these patients (*vide supra*). A shorter median follow-up of patients treated with demethylating agents, albeit with almost 200 patients, has very recently been provided by the MD Anderson group [162]. None of the patients observed after continued treatment with decitabine had developed secondary malignancy. Finally, chromosomal instability, which was noted in a mouse model engineered for severe hypomethylation (10% of normal methylation levels) was not observed in a retrospective analysis of the degree of chromosomal evolution in myelodysplasia patients followed with sequential cytogenetic studies [163]. The lack of tumorigenicity of long-term treatment with the HDAC inhibitor sodium phenolbutyrate in children with hereditary urea metabolic disorders also argues against gene reactivation in normal cells as a carcinogenic event.

In spite of the lack of evidence for mutagenic or carcinogenic effects of azanucleoside schedules at the present time, and in the patient populations studied long-term thus far, the development of DNA demethylating agents that are not incorporated in DNA is a very promising approach to more selectively target DNA methylation in cancer and hemoglobinopathies. The range of small molecules being developed for this approach includes nucleoside analogues such as zebularine [11], antisense oligonucleosides inhibiting DNA methylating activity, and, as recently reported [164], small interfering RNAs against Dnmt3b and Dnmt1. The preclinical efficacy of these different approaches is being determined in animal models of different cancers, and hopefully will allow more targeted therapy of hypermethylation in tumors in the future.

REFERENCES

1. D'Incalci, M. et al., DNA alkali-labile sites induced by incorporation of 5-aza-2'-deoxycytidine into DNA of mouse leukemia L1210 cells, *Cancer Res.*, 45(7), 3197–3202, 1985.
2. Li, L.H. et al., Cytotoxicity and mode of action of 5-azacytidine on L1210 leukemia, *Cancer Res.*, 30(11), 2760–2769, 1970.
3. Wilson, V.L., P.A. Jones, and R.L. Momparler, Inhibition of DNA methylation in L1210 leukemic cells by 5-aza-2'-deoxycytidine as a possible mechanism of chemotherapeutic action, *Cancer Res.*, 43(8), 3493–3496, 1983.
4. Hrodek, O. and J. Vesely, 5-azacytidine in childhood leukemia, *Neoplasma*, 18(5), 493–503, 1971.
5. Baehner, R.L. et al., Improved remission induction rate with D-ZAPO but unimproved remission duration with addition of immunotherapy to chemotherapy in previously untreated children with ANLL, *Med. Pediatr. Oncol.*, 7(2), 127–139, 1979.
6. Momparler, R.L., G.E. Rivard, and M. Gyger, Clinical trial on 5-aza-2'-deoxycytidine in patients with acute leukemia, *Pharmacol. Ther.*, 30(3), 277–286, 1985.
7. Taylor, S.M. and P.A. Jones, Multiple new phenotypes induced in 10T1/2 and 3T3 cells treated with 5-azacytidine, *Cell*, 17(4), 771–779, 1979.
8. Constantinides, P.G., P.A. Jones, and W. Gevers, Functional striated muscle cells from non-myoblast precursors following 5-azacytidine treatment, *Nature*, 267(5609), 364–366, 1977.
9. Konieczny, S.F. and C.P. Emerson, Jr., 5-Azacytidine induction of stable mesodermal stem cell lineages from 10T1/2 cells: evidence for regulatory genes controlling determination, *Cell*, 38(3), 791–800, 1984.
10. Jones, P.A. et al., *De novo* methylation of the MyoD1 CpG island during the establishment of immortal cell lines, *Proc. Natl. Acad. Sci. USA*, 187(16), 6117–6121, 1990.
11. Cheng, J.C. et al., Continuous zebularine treatment effectively sustains demethylation in human bladder cancer cells, *Mol. Cell Biol.*, 24(3), 1270–1278, 2004.
12. Segura-Pacheco, B. et al., Reactivation of tumor suppressor genes by the cardiovascular drugs hydralazine and procainamide and their potential use in cancer therapy, *Clin. Cancer Res.*, 9(5), 1596–1603, 2003.
13. Villar-Garea, A. et al., Procaine is a DNA-demethylating agent with growth-inhibitory effects in human cancer cells, *Cancer Res.*, 63(16), 4984–4989, 2003.
14. Fang, M.Z. et al., Tea polyphenol (-)-epigallocatechin-3-gallate inhibits DNA methyltransferase and reactivates methylation-silenced genes in cancer cell lines, *Cancer Res.*, 63(22), 7563–7570, 2003.
15. Jones, P.A. and S.M. Taylor, Cellular differentiation, cytidine analogs and DNA methylation, *Cell*, 20(1), 85–93, 1980.
16. Santini, V., H.M. Kantarjian, and J.P. Issa, Changes in DNA methylation in neoplasia: pathophysiology and therapeutic implications, *Ann. Intern. Med.*, 134(7), 573–586, 2001.
17. Jüttermann, R., E. Li, and R. Jaenisch, Toxicity of 5-aza-2'-deoxycytidine to mammalian cells is mediated primarily by covalent trapping of DNA methyltransferase rather than DNA demethylation, *Proc. Natl. Acad. Sci. USA*, 91(25), 11797–11801, 1994.
18. Woodcock, T.M. et al., Biochemical, pharmacological, and phase I clinical evaluation of pseudoisocytidine, *Cancer Res.*, 40(11), 4243–4249, 1980.

19. Cheng, J.C. et al., Inhibition of DNA methylation and reactivation of silenced genes by zebularine, *J. Natl. Cancer Inst.*, 95(5), 399–409, 2003.

20. Goldberg, R.M. et al., Phase I and pharmacological trial of fazarabine (Ara-AC) with granulocyte colony-stimulating factor, *Clin. Cancer Res.*, 3(12 Pt 1), 2363–2370, 1997.

21. Traganos, F. et al., Effects of dihydro-5-azacytidine on cell survival and cell cycle progression of cultured mammalian cells, *Cancer Res.*, 41(3), 780–789, 1981.

22. Von Hoff, D.D., M. Slavik, and F.M. Muggia, 5-Azacytidine. A new anticancer drug with effectiveness in acute myelogenous leukemia, *Ann. Intern. Med.*, 85(2), 237–245, 1975.

23. Glover, A.B. et al., Biochemistry of azacitidine: a review, *Cancer Treat. Rep.*, 71(10), 959–964, 1987.

24. Saiki, J.H. et al., Effect of schedule on activity and toxicity of 5-azacytidine in acute leukemia: a Southwest Oncology Group Study, *Cancer*, 47(7), 1739–1742, 1981.

25. Schwartsmann, G. et al., Decitabine (5-Aza-2′-deoxycytidine; DAC) plus daunorubicin as a first line treatment in patients with acute myeloid leukemia: preliminary observations, *Leukemia*, 11 Suppl. 1, S28–31, 1997.

26. Karon, M. et al., 5-Azacytidine: a new active agent for the treatment of acute leukemia, *Blood*, 42(3), 359–365, 1973.

27. McCredie, K.B. et al., Treatment of acute leukemia with 5-azacytidine (NSC-102816), *Cancer Chemother. Rep.*, 57(3), 319–323, 1973.

28. Saiki, J.H. et al., 5-Azacytidine in acute leukemia, *Cancer*, 42(5), 2111–2114, 1978.

29. Glover, A.B. et al., Azacitidine: 10 years later, *Cancer Treat. Rep.*, 71(7–8), 737–746, 1987.

30. Larson, R.A. et al., Response to 5-azacytidine in patients with refractory acute nonlymphocytic leukemia and association with chromosome findings, *Cancer*, 49(11), 2222–2225, 1982.

31. Goldberg, J. et al., Mitoxantrone and 5-azacytidine for refractory/relapsed ANLL or CML in blast crisis: a leukemia intergroup study, *Am. J. Hematol.*, 43(4), 286–290, 1993.

32. Van Echo, D.A., K.M. Lichtenfeld, and P.H. Wiernik, Vinblastine, 5-azacytidine, and VP-16-213 therapy for previously treated patients with acute nonlymphocytic leukemia, *Cancer Treat. Rep.*, 61(8), 1599–1602, 1977.

33. Omura, G.A. et al., Treatment of refractory adult acute leukemia with 5-azacytidine plus beta-2′-deoxythioguanosine, *Cancer Treat. Rep.*, 63(2), 209–210, 1979.

34. Martelo, O.J., G.O. Broun, Jr., and P.J. Petruska, Phase I study of pyrazofurin and 5-azacytidine in refractory adult acute leukemia, *Cancer Treat. Rep.*, 65(3–4), 237–239, 1981.

35. Baehner, R.L. et al., Contrasting benefits of two maintenance programs following identical induction in children with acute nonlymphocytic leukemia: a report from the Childrens Cancer Study Group, *Cancer Treat. Rep.*, 68(10), 1269–1272, 1984.

36. Vogler, W.R. et al., A randomized comparison of postremission therapy in acute myelogenous leukemia: a Southeastern Cancer Study Group trial, *Blood*, 63(5), 1039–1045, 1984.

37. Winton, E.F. et al., Sequentially administered 5-azacitidine and amsacrine in refractory adult acute leukemia: a phase I-II trial of the Southeastern Cancer Study Group, *Cancer Treat. Rep.*, 69(7–8), 807–811, 1985.

38. Dutcher, J.P. et al., Phase II study of mitoxantrone and 5-azacytidine for accelerated and blast crisis of chronic myelogenous leukemia: a study of the Eastern Cooperative Oncology Group, *Leukemia*, 6(8), 770–775, 1992.

39. Rivard, G.E. et al., Phase I study on 5-aza-2'-deoxycytidine in children with acute leukemia, *Leuk. Res.*, 5(6), 453–462, 1981.

40. Pinto, A. et al., 5-Aza-2'-deoxycytidine as a differentiation inducer in acute myeloid leukaemias and myelodysplastic syndromes of the elderly, *Bone Marrow Transplant,* 4 Suppl 3, 28–32, 1989.

41. Richel, D.J., L.P. Colly, and R. Willemze, The antileukaemic activity of 5-Aza-2-deoxycytidine (Aza-dC) in patients with relapsed acute leukaemia, *Bone Marrow Transplant.*, 4 Suppl 3, 64, 1989.

42. Petti, M.C. et al., Pilot study of 5-aza-2'-deoxycytidine (Decitabine) in the treatment of poor prognosis acute myelogenous leukemia patients: preliminary results, *Leukemia,* 7 Suppl 1, 36–41, 1993.

43. Willemze, R., E. Archimbaud, and P. Muus, Preliminary results with 5-aza-2'-deoxycytidine (DAC)-containing chemotherapy in patients with relapsed or refractory acute leukemia.The EORTC Leukemia Cooperative Group, *Leukemia*, 7 Suppl 1, 49–50, 1993.

44. Kantarjian, H.M. et al., Results of decitabine therapy in the accelerated and blastic phases of chronic myelogenous leukemia, *Leukemia*, 11(10), 1617–1620, 1997.

45. Willemze, R. et al., A randomized phase II study on the effects of 5-Aza-2'-deoxycytidine combined with either amsacrine or idarubicin in patients with relapsed acute leukemia: an EORTC Leukemia Cooperative Group phase II study (06893), *Leukemia*, 11 Suppl 1, S24–27, 1997.

46. Sacchi, S. et al., Chronic myelogenous leukemia in nonlymphoid blastic phase: analysis of the results of first salvage therapy with three different treatment approaches for 162 patients, *Cancer*, 86(12), 2632–2641, 1999.

47. Ravandi, F. et al., Decitabine with allogeneic peripheral blood stem cell transplantation in the therapy of leukemia relapse following a prior transplant: results of a phase I study, *Bone Marrow Transplant.*, 27(12), 1221–1225, 2001.

48. de Lima, M. et al., Long-term follow-up of a phase I study of high-dose decitabine, busulfan, and cyclophosphamide plus allogeneic transplantation for the treatment of patients with leukemias, *Cancer*, 97(5), 1242–1247, 2003.

49. Kantarjian, H.M. et al., Results of decitabine (5-aza-2'deoxycytidine) therapy in 130 patients with chronic myelogenous leukemia, *Cancer*, 98(3), 522–528, 2003.

50. Pinto, A. and V. Zagonel, 5-Aza-2'-deoxycytidine (Decitabine) and 5-azacytidine in the treatment of acute myeloid leukemias and myelodysplastic syndromes: past, present and future trends, *Leukemia*, 7 Suppl 1, 51–60, 1993.

51. Silverman, L.R. et al., Effects of treatment with 5-azacytidine on the *in vivo* and *in vitro* hematopoiesis in patients with myelodysplastic syndromes, *Leukemia*, 7 Suppl 1, 21–29, 1993.

52. Silverman, L.R. et al., Randomized controlled trial of azacitidine in patients with the myelodysplastic syndrome: a study of the cancer and leukemia group, *Clin. Oncol.*, 20(10), 2429–2440, 2002.

53. Greenberg, P. et al., International scoring system for evaluating prognosis in myelodysplastic syndromes, *Blood*, 89(6), 2079–2088, 1997.

54. Kornblith, A.B. et al., Impact of azacytidine on the quality of life of patients with myelodysplastic syndrome treated in a randomized phase III trial: a Cancer and Leukemia Group B study, *J. Clin. Oncol.*, 20(10), 2441–2452, 2002.

55. Wijermans, P.W. et al., Continuous infusion of low-dose 5-Aza-2'-deoxycytidine in elderly patients with high-risk myelodysplastic syndrome, *Leukemia*, 11 Suppl 1, S19–23, 1997.

56. Wijermans, P.W., M. Lübbert, and G. Verhoef, Low dose decitabine for elderly high risk MDS patients: who will respond? *Blood*, 100(11,1), 96a, 2002.

57. Lübbert, M. et al., Cytogenetic responses in high-risk myelodysplastic syndrome following low-dose treatment with the DNA methylation inhibitor 5-aza-2'-deoxycytidine, *Br. J. Haematol.*, 114(2), 349–357, 2001.

58. Issa, J.P. et al., Phase I study of low-dose prolonged exposure schedules of the hypomethylating agent 5-aza-2'-deoxycytidine (Decitabine) in hematopoietic malignancies, *Blood*, 103, 1635–1640, 2004.

59. Lee, E.J. et al., Low dose 5-azacytidine is ineffective for remission induction in patients with acute myeloid leukemia, *Leukemia*, 4(12), 835–838, 1990.

60. Chitambar, C.R. et al., Evaluation of continuous infusion low-dose 5-azacytidine in the treatment of myelodysplastic syndromes, *Am. J. Hematol.*, 37(2), 100–104, 1991.

61. Silverman, L.R. et al., Azacitidine (AzaC) in myelodysplastic syndromes (MDS), CALGB Studies 8421 and 8921, *Ann. Hematol.*, 68 (Suppl 2), 21a, 1994.

62. Rugo, H. et al., Compassionate use of subcutaneous 5-azacytidine (AzaC) in the treatment of myelodysplastic syndromes (MDS), *Leuk. Res.*, 23 (Suppl 1), 72, 1999.

63. Miller, C.B. et al., A phase I dose-descalation trial of combined DNA methyltransferase (MeT)/histone deacetylase (HDAC) inhibition in myeloid malignancies, *Blood*, 98 (Suppl 1), 622, 2001.

64. Silverman, L.R., Targeting hypomethylation of DNA to achieve cellular differentiation in myelodysplastic syndromes (MDS), *Oncologist*, 6 Suppl 5, 8–14, 2001.

65. Gryn, J. et al., Treatment of myelodysplastic syndromes with 5-azacytidine, *Leuk. Res.*, 26(10): p. 893–897, 2002.

66. Zagonel, V. et al., 5-Aza-2'-deoxycytidine (Decitabine) induces trilineage response in unfavourable myelodysplastic syndromes, *Leukemia*, 7 Suppl 1, 30–35, 1993.

67. Wijermans, P. et al., Low-dose 5-aza-2'-deoxycytidine, a DNA hypomethylating agent, for the treatment of high-risk myelodysplastic syndrome: a multicenter phase II study in elderly patients, *J. Clin. Oncol.*, 18(5), 956–962, 2000.

68. Lübbert, M. et al., Acute myeloid leukemia of the elderly and frail: hematologic and cytologic remissions attained by repeated courses of low-dose decitabine (DAC) and outpatient management, *Blood*, 100(11,2), 268b, 2002.

69. Rosato, R.R. and S. Grant, Histone deacetylase inhibitors in clinical development, *Expert Opin. Investig. Drugs*, 13(1), 21–38, 2004.

70. Cameron, E.E. et al., Synergy of demethylation and histone deacetylase inhibition in the re-expression of genes silenced in cancer, *Nat. Genet.*, 21(1), 103–107, 1999.

71. Belinsky, S.A. et al., Inhibition of DNA methylation and histone deacetylation prevents murine lung cancer, *Cancer Res.*, 63(21), 7089–7093, 2003.

72. Shaker, S. et al., Preclinical evaluation of antineoplastic activity of inhibitors of DNA methylation (5-aza-2'-deoxycytidine) and histone deacetylation (trichostatin A, depsipeptide) in combination against myeloid leukemic cells, *Leuk. Res.*, 27(5), 437–444, 2003.

73. Gore, S.D. et al., Impact of prolonged infusions of the putative differentiating agent sodium phenylbutyrate on myelodysplastic syndromes and acute myeloid leukemia, *Clin. Cancer Res.*, 8(4), 963–970, 2002.

74. Leone, G. et al., Inhibitors of DNA methylation in the treatment of hematological malignancies and MDS, *Clin. Immunol.*, 109(1), 89–102, 2003.

75. Camacho, L.H. et al., Transcription modulation: a pilot study of sodium phenylbutyrate plus 5-azacytidine, *Blood*, 98 (Suppl 1), 460, 2001.

76. Gojo, I. et al., Phase I study of histone deacetylase inhibitor (HDI) MS-275 in adults with refractory or relapsed hematologic malignancies, *Blood*, 102(11,1), 388a, 2003.

77. Kuendgen, A.M. et al., Valproic acid alone or in combination with all-trans-retinoic acid (ATRA) for the treatment of myelodysplastic syndromes and sAML/MDS, *Blood,* in press.

78. Momparler, R.L. et al., Pilot phase I-II study on 5-aza-2'-deoxycytidine (Decitabine) in patients with metastatic lung cancer, *Anticancer Drugs,* 8(4), 358–368, 1997.

79. Pohlmann, P. et al., Phase II trial of cisplatin plus decitabine, a new DNA hypo-methylating agent, in patients with advanced squamous cell carcinoma of the cervix, *Am. J. Clin. Oncol.,* 25(5), 496–501, 2002.

80. Abele, R. et al.,The EORTC Early Clinical Trials Cooperative Group experience with 5-aza-2'-deoxycytidine (NSC 127716) in patients with colorectal, head and neck, renal carcinomas and malignant melanomas, *Eur. J. Cancer Clin. Oncol.,* 23(12), 1921–1924, 1987.

81. Schwartsmann, G. et al., A phase I trial of cisplatin plus decitabine, a new DNA-hypomethylating agent, in patients with advanced solid tumors and a follow-up early phase II evaluation in patients with inoperable non-small cell lung cancer, *Invest. New Drugs,* 18(1), 83–91, 2000.

82. Curt, G.A. et al., A phase I and pharmacokinetic study of dihydro-5-azacytidine (NSC 264880), *Cancer Res.,* 45(7), 3359–3363, 1985.

83. Yogelzang, N.J. et al., Dihydro-5-azacytidine in malignant mesothelioma. A phase II trial demonstrating activity accompanied by cardiac toxicity. Cancer and Leukemia Group B, *Cancer,* 79(11), 2237–2242, 1997.

84. Moertel, C.G. et al., Phase II study of 5-azacytidine (NSC-102816) in the treatment of advanced gastrointestinal cancer, *Cancer Chemother. Rep.,* 56(5), 649–652, 1972.

85. Cunningham, T.J. et al., Comparison of 5-azacytidine (NSC-102816) with CCNU (NSC-79037) in the treatment of patients with breast cancer and evaluation of the subsequent use of cyclophosphamide (NSC-26271), *Cancer Chemother. Rep.,* 58(5 Pt 1), 677–681, 1974.

86. Bellet, R.E. et al., Clinical trial with subcutaneously administered 5-azacytidine (NSC-102816), *Cancer Chemother. Rep.,* 58(2), 217–222, 1974.

87. Sessa, C. et al., Phase II study of 5-aza-2'-deoxycytidine in advanced ovarian carci-noma. The EORTC Early Clinical Trials Group, *Eur. J. Cancer,* 26(2), 137–138, 1990.

88. Vermorken, J.B. et al., 5-aza-2'-deoxycytidine in advanced or recurrent cancer of the uterine cervix, *Eur. J. Cancer,* 27(2), 216–217, 1991.

89. Thibault, A. et al., A phase II study of 5-aza-2'deoxycytidine (decitabine) in hormone independent metastatic (D2) prostate cancer, *Tumori,* 84(1), 87–89, 1998.

90. Bellet, R.E. et al., Phase II study of subcutaneously administered 5-azacytidine (NSC-102816) in patients with metastatic malignant melanoma, *Med. Pediatr. Oncol.,* 4(1), 11–15, 1978.

91. Creagan, E.T. et al., A phase II study of 5,6-dihydro-5-azacytidine hydrochloride in disseminated malignant melanoma, *Am. J. Clin. Oncol.,* 16(3), 243–244, 1993.

92. Weiss, A.J. et al., Phase II study of 5-azacytidine in solid tumors, *Cancer Treat. Rep.,* 61(1), 55–58, 1977.

93. van Groeningen, C.J. et al., Phase I and pharmacokinetic study of 5-aza-2'-deoxycy-tidine (NSC 127716) in cancer patients, *Cancer Res.,* 46(9), 4831–4836, 1986.

94. Aparicio, A. et al., Phase I trial of continuous infusion 5-aza-2'-deoxycytidine., *Cancer Chemother. Pharmacol.,* 51(3), 231–239, 2002.

95. van der Ploeg, L.H. and R.A. Flavell, DNA methylation in the human gamma delta beta-globin locus in erythroid and nonerythroid tissues, *Cell,* 19(4), 947–958, 1980.

96. DeSimone, J. et al., 5-Azacytidine stimulates fetal hemoglobin synthesis in anemic baboons, *Proc. Natl. Acad. Sci. USA,* 79(14), 4428–4431, 1982.

97. Ley, T.J. et al., 5-azacytidine selectively increases gamma-globin synthesis in a patient with beta+ thalassemia, *N. Engl. J. Med.*, 307(24), 1469–1475, 1982.

98. Charache, S. et al., Treatment of sickle cell anemia with 5-azacytidine results in increased fetal hemoglobin production and is associated with nonrandom hypomethylation of DNA around the gamma-delta-beta-globin gene complex, *Proc. Natl. Acad. Sci. USA*, 80(15), 4842–4846, 1983.

99. Lowrey, C.H. and A.W. Nienhuis, Brief report: treatment with azacitidine of patients with end-stage beta-thalassemia, *N. Engl. J. Med.*, 329(12), 845–848, 1993.

100. Koshy, H. et al., Augmentation of fetal hemoglobin (HbF) levels by low dose short duration 5-aza-2′-deoxycytidine (decitabine) administration in sickle cell anemia patients who had no HbF elevation following hydroxyurea therapy, *Blood*, 92(Suppl 1), 306b, 1998.

101. Saunthararajah, Y. et al., Effects of 5-aza-2′-deoxycytidine on fetal hemoglobin levels, red cell adhesion, and hematopoietic differentiation in patients with sickle cell disease, *Blood*, 102(12), 3865–3970, 2003

102. Ley, T.J. et al., 5-Azacytidine increases gamma-globin synthesis and reduces the proportion of dense cells in patients with sickle cell anemia, *Blood*, 62(2), 370–380, 1983.

103. Dover, G.J. et al., 5-Azacytidine increases HbF production and reduces anemia in sickle cell disease: dose-response analysis of subcutaneous and oral dosage regimens, *Blood*, 66(3), 527–532, 1985.

104. Humphries, R.K. et al., 5-Azacytidine acts directly on both erythroid precursors and progenitors to increase production of fetal hemoglobin, *J. Clin. Invest.*, 75(2), 547–557, 1985.

105. Dunbar, C. et al., 5-Azacytidine treatment in a beta (0)-thalassaemic patient unable to be transfused due to multiple alloantibodies, *Br. J. Haematol.*, 72(3), 467–468, 1989.

106. Koshy, M. et al., 2-deoxy 5-azacytidine and fetal hemoglobin induction in sickle cell anemia, *Blood*, 96(7), 2379–2384, 2000.

107. DeSimone, J. et al., Maintenance of elevated fetal hemoglobin levels by decitabine during dose interval treatment of sickle cell anemia, *Blood*, 99(11), 3905–3008, 2002.

108. el-Deiry, W.S. et al., WAF1, a potential mediator of p53 tumor suppression, *Cell*, 75(4), 817–825, 1993.

109. Xiong, Y. et al., p21 is a universal inhibitor of cyclin kinases, *Nature*, 366(6456), 701–704, 1993.

110. Polyak, K. et al., p27Kip1, a cyclin-CDK inhibitor, links transforming growth factor-beta and contact inhibition to cell cycle arrest, *Genes Dev.*, 8(1): p. 9-22, 1994.

111. Toyoshima, H. and T. Hunter, p27, a novel inhibitor of G1 cyclin-CDK protein kinase activity, is related to p21. *Cell*, 78(1), 67–74, 1994.

112. Lee, M.H., I. Reynisdottir, and J. Massague, Cloning of p57KIP2, a cyclin-dependent kinase inhibitor with unique domain structure and tissue distribution, *Genes Dev.*, 9(6), 639–649, 1995.

113. Serrano, M., G.J. Hannon, and D. Beach, A new regulatory motif in cell-cycle control causing specific inhibition of cyclin D/CDK4, *Nature*, 366(6456), 704–707, 1993.

114. Hannon, G.J. and D. Beach, p15INK4B is a potential effector of TGF-beta-induced cell cycle arrest, *Nature*, 371(6494), 257–261, 1994.

115. Guan, K.L. et al., Growth suppression by p18, a p16INK4/MTS1- and p14INK4B/MTS2-related CDK6 inhibitor, correlates with wild-type pRb function, *Genes Dev.*, 8(24), 2939–2952, 1994.

116. Guan, K.L. et al., Isolation and characterization of p19INK4d, a p16-related inhibitor specific to CDK6 and CDK4, *Mol. Biol. Cell*, 7(1), 57–70, 1996.

117. Chan, F.K. et al., Identification of human and mouse p19, a novel CDK4 and CDK6 inhibitor with homology to p16ink4, *Mol. Cell Biol.*, 15(5), 2682–2688, 1995.

118. Baylin, S.B. et al., Alterations in DNA methylation: a fundamental aspect of neoplasia. *Adv. Cancer Res.*, 72, 141–196, 1998.

119. Esteller, M., CpG island hypermethylation and tumor suppressor genes: a booming present, a brighter future, *Oncogene*, 21(35), 5427–5440, 2001.

120. Claus, R. and M. Lübbert, Epigenetic targets in hematopoietic malignancies, *Oncogene*, 22(42), 6489–6496, 2003.

121. Merlo, A. et al., 5′ CpG island methylation is associated with transcriptional silencing of the tumour suppressor p16/CDKN2/MTS1 in human cancers, *Nat. Med.*, 1(7), 686–692, 1995.

122. Sanchez-Aguilera, A. et al., Silencing of the p18INK4c gene by promoter hypermethylation in the Reed-Sternberg cells in Hodgkin's lymphomas, *Blood*, 103, 2351–2357, 2004.

123. Nakatsuka, S. et al., Methylation of promoter region in p27 gene plays a role in the development of lymphoid malignancies, *Int. J. Oncol.*, 22(3), 561–568, 2003.

124. Li, Y. et al., Aberrant DNA methylation of p57(KIP2) gene in the promoter region in lymphoid malignancies of B-cell phenotype, *Blood*, 100(7), 2572–2577, 2002.

125. Otterson, G.A. et al., CDKN2 gene silencing in lung cancer by DNA hypermethylation and kinetics of p16INK4 protein induction by 5-aza-2′deoxycytidine, *Oncogene*, 11(6), 1211–1216, 1995.

126. Scandura, J.M., D. McGrogan, and S.D. Nimer, Frequent alterations in the cyclin-dependent kinase inhibitor p57KIP2 in AML-derived cell lines: association with loss of TGF beta responsiveness, *Blood*, 102, 2003 (abs.).

127. Konishi, N. et al., Heterogeneous methylation and deletion patterns of the INK4a/ARF locus within prostate carcinomas, *Am. J. Pathol.*, 160(4), 1207–1214, 2002.

128. Robertson, K.D. and P.A. Jones, The human ARF cell cycle regulatory gene promoter is a CpG island which can be silenced by DNA methylation and down-regulated by wild-type p53, *Mol. Cell Biol.*, 18(11), 6457–6473, 1998.

129. Dodge, J.E., A.F. List, and B.W. Futscher, Selective variegated methylation of the p15 CpG island in acute myeloid leukemia, *Intl. J. Cancer*, 78(5), 561–567, 1998.

130. Dodge, J.E., C. Munson, and A.F. List, KG-1 and KG-1a model the p15 CpG island methylation observed in acute myeloid leukemia patients, *Leuk. Res.*, 25(10), 917–925, 2001.

131. Issa, J.P., Decitabine, *Curr. Opin. Oncol.*, 15(6), 446–451, 2003.

132. Lübbert, M., DNA methylation inhibitors in the treatment of leukemias, myelodysplastic syndromes and hemoglobinopathies: clinical results and possible mechanisms of action, *Curr. Top. Microbiol. Immunol.*, 249, 135–164, 2000.

133. Pinto, A. et al., 5-Aza-2′-deoxycytidine induces terminal differentiation of leukemic blasts from patients with acute myeloid leukemias, *Blood*, 64(4), 922–929, 1984.

134. Nieto, M. et al., The absence of p53 is critical for the induction of apoptosis by 5-aza-2′-deoxycytidine, *Oncogene*, 23(3), 735–743, 2004.

135. Karpf, A.R. et al., Activation of the p53 DNA damage response pathway after inhibition of DNA methyltransferase by 5-aza-2′-deoxycytidine, *Mol. Pharmacol.*, 59(4), 751–757, 2001.

136. Zheng, S. et al., Correlations of partial and extensive methylation at the p14(ARF) locus with reduced mRNA expression in colorectal cancer cell lines and clinicopathological features in primary tumors, *Carcinogenesis*, 21(11), 2057–2064, 2000.

137. Latres, E. et al., Limited overlapping roles of P15(INK4b) and P18(INK4c) cell cycle inhibitors in proliferation and tumorigenesis, *Embo. J.*, 19(13), 3496–3506, 2000.

138. Wolff, L. et al., Hypermethylation of the Ink4b locus in murine myeloid leukemia and increased susceptibility to leukemia in p15(Ink4b)-deficient mice, *Oncogene*, 22(58), 9265–9274, 2003.

139. Daskalakis, M. et al., Demethylation of a hypermethylated P15/INK4B gene in patients with myelodysplastic syndrome by 5-aza-2′-deoxycytidine (decitabine) treatment, *Blood*, 100(8), 2957–2964, 2002.

140. Timmermann, S., P.W. Hinds, and K. Münger, Re-expression of endogenous p16ink4a in oral squamous cell carcinoma lines by 5-aza-2′-deoxycytidine treatment induces a senescence-like state, *Oncogene*, 17(26), 3445–3453, 1998.

141. Lavelle, D. et al., Decitabine induces cell cycle arrest at the G1 phase via p21(WAF1) and the G2/M phase via the p38 MAP kinase pathway, *Leuk. Res.*, 27(11), 999–1007, 2003.

142. Zhu, W.G. et al., Increased expression of unmethylated CDKN2D by 5-aza-2′-deoxycytidine in human lung cancer cells, *Oncogene*, 20(53), 7787–7796, 2001.

143. Jiemjit, A. et al., Impact of 5-azacytidine and 2-deoxy-5-azacytidine in histone acetylation and expression of the non-methylated p21WAF1/CIP1 gene, *Blood*, 102(11, 2), 213b, 2003.

144. Shin, J.Y. et al., Mechanism for inactivation of the KIP family cyclin-dependent kinase inhibitor genes in gastric cancer cells, *Cancer Res.*, 60(2), 262–265, 2000.

145. Qian, X. et al., DNA methylation regulates p27kip1 expression in rodent pituitary cell lines. *Am. J. Pathol.*, 153(5), 1475–1482, 1998.

146. Garber, K., Silence of the genes: cancer epigenetics arrives, *J. Natl. Cancer Inst.*, 94(11), 793–795, 2002.

147. Marks, P.A. et al., Histone deacetylase inhibitors as new cancer drugs, *Curr. Opin. Oncol.*, 13(6), 477–483, 2001

148. Nguyen, C.T. et al., Histone H3-lysine 9 methylation is associated with aberrant gene silencing in cancer cells and is rapidly reversed by 5-aza-2′-deoxycytidine, *Cancer Res.*, 62(22), 6456–6461, 2002.

149. Zhu, W.G. et al., DNA methyltransferase inhibition enhances apoptosis induced by histone deacetylase inhibitors, *Cancer Res.*, 61(4), 1327–1333, 2001.

150. Esteller, M. et al., p14ARF silencing by promoter hypermethylation mediates abnormal intracellular localization of MDM2, *Cancer Res.*, 61(7), 2816–2821, 2001.

151. Richel, D.J. et al., The antileukaemic activity of 5-Aza-2 deoxycytidine (Aza-dC) in patients with relapsed and resistant leukaemia, *Br. J. Cancer*, 64(1), 144–148, 1991.

152. Sigalotti, L. et al., 5-Aza-2′-deoxycytidine (decitabine) treatment of hematopoietic malignancies: a multimechanism therapeutic approach? *Blood*, 101(11), 4644–4646; discussion 4645–4646, 2003.

153. Baylin, S.B., Mechanisms underlying epigenetically mediated gene silencing in cancer, *Semin. Cancer Biol.*, 12(5), 331–337, 2002.

154. El-Osta, A., The rise and fall of genomic methylation in cancer, *Leukemia*, 18(2), 233–237, 2004.

155. Plumb, J.A. et al., Reversal of drug resistance in human tumor xenografts by 2′-deoxy-5-azacytidine-induced demethylation of the hMLH1 gene promoter, *Cancer Res.*, 60(21), 6039–6044, 2000.

156. Lübbert, M. et al., Multiple hypermethylated genes are potential *in vivo* targets of demethylating agents, *Blood*, 101(11), 4645–4646, 2003.

157. Laird, P.W. et al., Suppression of intestinal neoplasia by DNA hypomethylation, *Cell*, 81(2), 197–205, 1995.

158. Trinh, B.N. et al., DNA methyltransferase deficiency modifies cancer susceptibility in mice lacking DNA mismatch repair., *Mol. Cell Biol.*, 22(9), 2906–2917, 2002.

159. Xu, G.L. et al., Chromosome instability and immunodeficiency syndrome caused by mutations in a DNA methyltransferase gene, *Nature*, 402(6758), 187–191, 1999.

160. Hansen, R.S. et al., The DNMT3B DNA methyltransferase gene is mutated in the ICF immunodeficiency syndrome. *Proc. Natl. Acad. Sci. USA*, 96(25), 14412–14417, 1999.

161. Brown, D.C. et al., ICF syndrome (immunodeficiency, centromeric instability and facial anomalies): investigation of heterochromatin abnormalities and review of clinical outcome, *Hum. Genet.*, 96(4), 411–416, 1995.

162. Yang, A.S. et al., Comment on "Chromosomal instability and tumors promoted by DNA hypomethylation" and "Induction of tumors in nice by genomic hypomethylation." *Science*, 302(5648), 1153; author reply 1153, 2003.

163. Lübbert, M., P.W. Wijermans, and G. Verhoef, Karyotypic stability or chromosomal evolution during treatment with a DNA demethylating agent? a serial study of patients with the myelodysplastic syndrome, *Blood*, 102(11,1), 427a, 2003.

164. Leu, Y.W. et al., Double RNA interference of DNMT3b and DNMT1 enhances DNA demethylation and gene reactivation, *Cancer Res.*, 63(19), 6110–6115, 2003.

Index

A

ABO genes, 4
Acetic acid, 125
Acetone, 125
Acute lymphoblastic leukemia (ALL), 184
Acute myeloid leukemia (AML), 185
    aberrant methylation in, 102
    azanucleoside therapy, 186, 190–192
    combination therapy, 193–194
Adriamycin, 161
AFLP, 86
Aging, DNA methylation and, 14, 46
Agouti, 46
AIMS, 6, 86
    methyl-enriched, 90–92
AKT2, 103
Albright hereditary osteodystrophy, 5
Alkaline phosphatase, 115, 117
Alpha-thalassemia mental retardation (ATR-X)
         syndrome, 4, 45
Alu elements, 2
    reactivation of, 4
Alu-PCR, 85
Alzheimer's disease, 5
Amplicons, 74, 79, 80
Amplification
    of bisulfite-converted DNA, 57–59
    in ChIP assays, 131
    chromosomal regions, 103
    methylation-specific, 66
    nonselective, 66
    of undigested DNA, 81–82
Amplification of intermethylated sites (AIMS),
         6, 86, 90–92
Amplified fragment length polymorphism
         (AFLP), 86
Aneuploidy, 4
Angelman syndrome, 5, 39–40, 98
Antibodies
    anti-methylcytosine, 114
    in ChIP assays, 131
    specificity of, 122–123
Antigens, fixation protocols, 123–126
Anti-phospho-Ser, 122
Antisense oligonucleotides, 172, 203
    inhibitors, 156

Apaf-1, 161
APC, cancer-associated methylation, 13–16, 20
Apc$^{Min/+}$, 142–143
Apoptosis, 201
    reactivation of, 159, 161, 170
AP-PCR, 85–92
Arabinofuranosyl-5-azacytosine, 155
Arbitrarily primed PCR, 85–92
Array comparative genome hybridization, 103
*Asc*I, 96
*Asc*I–*Eco*RV boundary library, 105
Atherosclerotic disease, 5
ATP10C gene, 40
ATR-X syndrome, 4, 45
ATRX, 4, 45
Autoimmunity, and DNA methylation, 4–5
5-Aza-2′-deoxycytidine, 118, 143, 152-156, 184;
       see also Decitabine
    gene reactivation by, 3
    high-dose, 186
    and interferon, 161–162
    low-dose, 192
    nonspecificity of, 159
    sensitization of cancer cells, 161
    and trichostatin A, 161
5-Azacytidine, 42, 152-156, 171–172, 184
    analogs, 175
    clinical trials, 184–185
    high-dose, 185–186, 202
    long-term, 203
    low-dose, 190–192, 202, 203
    and sodium phenylbutyrate, 192
    treatment of solid tumors, 194, 202
5-Azacytosine, analogs, 172
Azanucleosides, 152–156
    clinical development of, 184–185
    in hemoglobinopathies, 196–198
    and inappropriate gene reactivation, 202–203
    low-dose chemotherapy, 190–194, 202

B

Barr body, 29
B-cell lymphomas, 20, 144
BCNU, 2
Beckwith-Wiedemann syndrome, 5, 40–41, 98

213